Henry Worms

The Earth and Its Mechanism :

Being an Account of the Various Proofs of the Rotation of the Earth

Henry Worms

The Earth and Its Mechanism :
Being an Account of the Various Proofs of the Rotation of the Earth

ISBN/EAN: 9783337003043

Printed in Europe, USA, Canada, Australia, Japan

Cover: Foto ©berggeist007 / pixelio.de

More available books at **www.hansebooks.com**

THE

EARTH AND ITS MECHANISM:

BEING AN ACCOUNT OF THE VARIOUS

PROOFS OF THE ROTATION OF THE EARTH.

WITH A DESCRIPTION OF THE INSTRUMENTS
USED IN THE EXPERIMENTAL DEMONSTRATIONS: TO WHICH IS ADDED THE THEORY
OF FOUCAULT'S PENDULUM AND GYROSCOPE.

BY HENRY WORMS, F.R.A.S., F.G.S.

'E pur si muove.'

WITH NUMEROUS WOODCUTS AND DIAGRAMS

LONDON:
LONGMAN, GREEN, LONGMAN, ROBERTS, & GREEN.
1862

TO

PROFESSOR WHEATSTONE, LL.D., F.R.S., &c.

WITH FEELINGS OF

THE DEEPEST RESPECT, AND IN

ADMIRATION OF THAT FERTILE GENIUS

WHICH HAS PRODUCED SO MANY USEFUL INVENTIONS IN SCIENCE

AND ART, THIS TREATISE IS INSCRIBED

BY HIS OBEDIENT SERVANT

THE AUTHOR.

PREFACE.

The present work consists of two parts, the one popular and adapted to the comprehension of the general reader; the other analytical and designed for the perusal of the mathematical student. The table of contents will sufficiently exhibit the plan of the whole as a complete collection of the various demonstrations of the Earth's rotation and the theorems connected therewith.

The principal object, however, has been to promulgate a knowledge of the more recent discoveries and experiments relating to this branch of science.

In the first instance, therefore, particular attention is drawn to Arago's application of Wheatstone's revolving mirror for the determination of the finite velocity of light, which not only affords evidence of the motion of translation as thought hitherto, but also of that of rotation.

Then a minute account is given of the latest experiments on falling bodies made by Professor Reich in the mines at Freiberg in addition to those of

Benzenberg and Guglielmini. They deserve special notice because their results are trustworthy, and the idea of undertaking them for ascertaining the rotation of the Earth originated with the illustrious Newton.

The Author has further endeavoured to give a clear and distinct idea of the phenomena exhibited by Foucault's pendulum and gyroscope, inasmuch as they furnish the most remarkable and direct proofs of the Earth's rotation.

In order to elucidate the principle on which the apparent change of the plane of oscillation depends (without having recourse to abstruse reasoning and calculation), the Author has made use of a number of auxiliary instruments originally invented for this purpose. A full description of these instruments in connection with that of the main apparatus is here offered to the public, accompanied throughout with corresponding graphical illustrations.

Finally, new modes for the suspension of the pendulum are proposed as improvements, modifications of the gyroscope with practical directions for using them are pointed out, and the conical motion of the Earth's axis itself, causing the precession of the equinoxes, is explained by Maxwell's dynamical top and Burr's apparatus.

With regard to the theoretical portion of the work containing the solution of intricate problems of analytical mechanics, and intended for the mathematical student, it is necessary to state that in the theory of the motion of the pendulum taken from 'Hansen's Theorie der Pendelbewegung,' the Author has allowed himself some omissions and additions, for which he is responsible; at the same time he ventures to express a hope that the Reader will be lenient in his criticism of the imperfections and shortcomings of this and other parts of the work, and take into consideration the extreme difficulty and abstruseness of the several subjects he has endeavoured to treat.

Goldsmith Building, Temple:
October 1862.

CONTENTS.

PART I.

INTRODUCTORY CHAPTER

DID THE ANCIENTS KNOW THE COPERNICAN SYSTEM?

CHAPTER I.

THE DISCOVERY OF THE EARTH'S MOTION.

CHAPTER II.

THE PROBABILITIES OF THE EARTH'S MOTION.

CHAPTER III.

PROOF OF THE EARTH'S ROTATION BY THE TRANSMISSION OF LIGHT.

CHAPTER IV.

THE EARTH'S ROTATION DEMONSTRATED BY THE VARYING FORCE OF GRAVITY IN DIFFERENT LATITUDES.

CHAPTER V.

THE DESCENT OF BODIES FROM GREAT HEIGHTS.

CHAPTER VI.

ON THE APPARENT CHANGE OF THE PLANE OF OSCILLATION OF THE FREE PENDULUM.

PART II.

CHAPTER I.

CHAPTER II.

CHAPTER III.

CHAPTER IV.

Note—Page 167.

PART I.

INTRODUCTORY CHAPTER.

DID THE ANCIENTS KNOW THE COPERNICAN SYSTEM?

ATTEMPTS have been made at various periods to show that Copernicus borrowed his ideas from the Pythagoreans, and he has therefore been called by Bailly the restorer of the true system of the world. The object which we have in view in this introduction is to prove that the ancients did not possess that amount of knowledge which would have enabled them to explain the motion of the celestial bodies in the manner which is now universally recognised. Before citing all the passages relating to the subject which are to be met with in the classics, we will quote those which are admitted by Copernicus, in his preface to the 'Revolutionibus Orbium Celestium,' to have led him in some degree to his discoveries. He says, 'Ac reperi quidem apud Ciceronem primum: Nicetam sensisse terram moveri; postea et apud Plutarchum inveni quosdam alios in ea fuisse opinione. Inde igitur occasionem nactus, cœpi ut ego de terræ mobilitate cogitare.' The two passages to which he alludes are the following. Cicero having spoken (Quæst. Acad. 2) of men who believe in the existence of antipodes, continues thus: 'Theophrastus narrates that Nicetas of Syracuse was of opinion that the sun, moon, and stars were at rest, and the earth alone moved, turning about its axis, by which the same phenomena were produced, as if the contrary were the case.' From this it is evident that either Cicero misinterpreted Theophrastes, or the latter did not understand Nicetas, or, what is still more probable, Nicetas himself had but a very imperfect knowledge of the subject, for it is impossible to explain the apparent movements of the sun, the moon, and other planets by the rotation of the earth alone. It is difficult to determine whether Copernicus meant this passage of Cicero, or whether he did not rather allude to the one in Lib. I. Tuscul., which briefly states, 'Primus Nicetas Syracusanus terram moveri,' etc. These five words might have led us to believe that Nicetas had a correct idea of the universe, were we not aware of the passage above cited. We may here aptly apply the old rule of the oracle, 'He who knows but little of a subject should be sparing of his words or hold his tongue. He who veils his meaning by the obscurity of his speech defies contradiction.'

The passage in Plutarch referred to by Copernicus is in the third Book, 'De Placitis Philosophorum,' caps. 11 and 13. 'The followers of Thales placed the earth in the centre of the universe; Xenophanes was the first who believed it to be in infinite space. The doctrine of Philolaus the Pythagorean was, that fire was the centre and focus of the universe; while the other Pythagoreans maintained that the earth was at rest, he asserted that it revolved in an oblique circle about a central fire like the sun and moon.' 'Ceteri terram manere dixerunt. Philolaus vero Pythagoræus, circum ignem verti obliquo in circulo similiter soli et lunæ.' 'Heraclides of Pontus, and Ecphantes the Pythagorean, believed that the earth had no translation in space, but revolved like a wheel upon its axis from west to east.' We gather from these quotations that mention is made of a motion of the earth, but not of the true one, and the following fact proves that no notice of it was contained in the writings of Philolaus. Plato put this question to the astronomers, 'How is the retrograde motion of the planets to be explained?' Eudoxus solved the problem by means of concentric spheres. The great Plato undoubtedly saw that this explanation was very complicated and far-fetched, and as he was in possession of Philolaus' manuscripts, having purchased them of his heirs for a hundred Mines,* it is very unlikely that he neglected to read them, especially as he held the author in such high esteem that he travelled to Italy on purpose to see him, which was in those days no small undertaking. We learn, therefore, from Eudoxus that the Pythagoreans did not know the Copernican system, otherwise he would never have introduced so complicated a combination of concentric circles in order to explain the celestial phenomena. In fact, none of the notions of the Greeks were founded on a true knowledge of astronomy, or upon a comparison of facts; everything was speculation without regard to inductive reasoning, and they consequently arrived at contradictory results. Some considered the earth to be a cube, others a mountain, raising its lofty and rounded summit from the midst of the universe.

Besides the passages quoted by Copernicus, and which we have just criticised, there are others in the writings of antiquity in which mention is made of a motion of the earth. Aristotle, in his second Book, 'De Cœlo,' says, 'Most people place the earth in the middle, but some

* A 'Mine' was worth about £4 1*s.* 3*d.*

who live in Italy and are called Pythagoreans assert that fire is the centre, and the earth is only a star, which, by its revolution, produces day and night.' At first sight this passage might lead us to infer that they were aware of one of the earth's motions, but examining closely their mode of investigation, we find that this isolated truth was not the result of observation, but an hypothesis based on illusory premises; for they reasoned in the following way — the noblest body should have the noblest place. Fire, in their estimation, ranked higher than earth, therefore to the former they assigned the 'place d'honneur.' 'Ignem hanc habere regionem dixerunt, quem Jovis carcerem sive custodiam nominant.'

In the 'Arenario' of Archimedes, it is stated that Aristarchus of Samos was of opinion that the stars moved and the sun was fixed, that the earth revolved in a circle round the latter, and that the sphere which carried the stars round rotated also about it. Aristarchus lived in the year 280 B.C., and the superstitious Athenians believed that his doctrines were contrary to religion, because they deprived the earth of its central position. Cleanthus, a disciple of Zeno, indicted him for despising the glory of the goddess Vesta. This prosecution by the Athenian inquisition is assigned by many as the reason that the imperfect knowledge of the true system was kept so secret as now to be lost altogether.

Athenian power, however, did not extend far, and the philosophers in other parts of Greece, in Asia Minor, and in Italy, where civilization was quite as high, could propagate their opinions without fear of inquisitorial persecution. This much we may admit on the testimony of Archimedes, that the opinions of Aristarchus excelled in clearness and correctness those of all the other Greeks; but they still did not possess any great astronomical importance, as they were not employed to explain the motion of the planets in their direct or retrograde motion. The whole planetary theory was unknown to him, and he, more Pythagorean than astronomer, never conceived that so complicated a mechanism was capable of so simple an explanation; and had he even imagined it, he never could have established it, owing to the total want of observations at that period. But the strongest proof that the true system was not known, is that no allusion is made to it by the greatest astronomers of that time. If the assertions made by the Greeks about the motion of the earth and the place of the sun were more than mere philosophical surmises, then Hipparchus would have availed himself of them for his own observations. In order to avoid the introduction of epicycles, he refrained from considering the earth as the centre of the system, but did not go so far as to assume its annual and diurnal motion. He calculated the first astronomical tables which served as a model for future ages, but he did not base them upon the Copernican hypothesis, though he might have done so without revealing the truth to the uninitiated, and neither Greek nor Roman priests could have discovered those points which they considered heretical. He despaired of explaining the stationary position or retrograde motion of the planets, and contented himself with collecting the results of observations previously made and comparing them with his own, in order to bequeath them to posterity. His successor Ptolemy is known as the champion of the principle of the earth's immobility, and we shall hereafter notice a few of his arguments. Having thus hastily glanced over the opinions of the Greeks, we will turn our attention to those of the Romans. We find that Seneca, their greatest philosopher, speaks several times in his 'Quæst. Nat.' of a motion of the earth. In the second chapter he says: 'It seems that the comets have something in common with the planets; therefore, if all the stars are worlds, the comets must be considered as such. But if the stars are mere elementary fire, which can last for six months without being extinguished by the rapid movement, then comets may also consist of this thin and subtile material without being endangered by the revolution of the firmament. This inquiry will also serve to show us whether the earth is at rest and the firmament moves round it, or the latter is at rest and the former turns. There were men who maintained that the heavens did not move but we did. This subject is worthy of investigation, in order to learn whether our habitation is at rest or endowed with rapid motion, whether God causes us to move, or merely the celestial bodies around us.'

In the 25th and 26th caps. of the same book, we find:—'We have so little knowledge of nature, that we must not feel surprised at not understanding the laws of the comets which only appear occasionally. There are even now nations which do not know why the moon has phases and is sometimes eclipsed, nor is it long since we found out the cause of these things. The time will come when that which is now hidden will be brought to light by the industry and intelligence of future generations. A lifetime is not sufficient to analyse the laws which regulate these phenomena, and are not the few years of our existence improperly divided between our studies and our passions? A time will come when our descendants will look back with astonishment and wonder at our obtuseness in understanding things which were so clear and obvious. Of the five stars which attract our attention from their continual change of position, we are at a loss to know which rises in the morning and which in the evening, whether they are in the direct course, and why they become retrograde. Some wise men say: you are mistaken in maintaining that a star is stationary or goes back, for celestial bodies must go forward in a fixed direction. Their path will always be the same in eternity, the movements of the universe are immutable; were they to cease, the world would sink into ruins.'

Seneca continues, Cap. 27:

'What is the reason that some of the stars appear to go backwards? It is because the sun moves towards them with such rapidity that their motion is relatively slower, and then their circular orbits are in such position to each other as to deceive those who look upon them; on the same principle that a ship in full sail sometimes appears to be at rest.'

We gather from these passages that the want was felt of establishing a simpler system of astronomy, and that a slight conception existed of the manner in which it might be accomplished; but Seneca, whose genius excelled that of all his contemporaries, had but an obscure idea of the earth's true motion, and was even unable to apply the scanty knowledge he possessed to solve the great problem of the mechanism of the universe.

Before we decide the question which is the object of this chapter, it is necessary to define the three essentially different points of the Copernican system.

1. The diurnal motion from west to east.

2. The annual motion around the sun.

3. The motion of the planets in their direct and retrograde courses.

With regard to the first, we have seen that Heraclides and Ecphantes were aware that by it the phenomena of day and night could be explained. Aristarchus of Samos had a notion of the second, and most of the Pythagoreans placed the sun, or fire (two expressions which they often confounded), in the centre of the universe, for the absurd reason we have stated above.

But the third, the most important of all, was completely unknown to the ancients, they did not possess the requisite astronomical observations. How trifling a knowledge of astronomy the Greeks had, and how utterly at a loss they were to invent a system, we gather from Plutarch's 'De Placitis Philosophorum,' in which propositions are laid down which prove nothing but the infantine state of science at that period.

Anaximander says: 'The figure of the earth is that of a column, and the sun and moon are great vessels filled with fire; at the top of them is an opening through which the fire escapes, but should the aperture be closed an eclipse takes place.' Anaximenes believed the heavens to be solid, and to consist of a kind of fine earthenware, and the stars to be like golden nails driven through it.

Anaxagoras considered that the solstices were caused by the air at the poles being so thick that the sun could not pass through. He also thought that the sun and moon were rocks cast off by centrifugal force, and ignited in the upper regions of fire.

Aristotle was of opinion that the spots on the moon were the reflected images of the seas and lands of the earth. Theophrastes believed that the celestial hemispheres were knitted together by means of the Milky Way, but so carelessly that spaces were left through which the fiery heavens could be seen. Xenophanes' doctrine was that the sun consisted of particles of fire extracted from the humid vapours of the earth, which were extinguished every night and renewed every morning, and that the stars were fiery clouds collected in the higher regions. This simple notion is expressed by Lucretius, when he asks:

> Unde mare, ingenui fontes æternaque longe
> Flumina suppeditant? Unde æther sidera pascit?

Summing up impartially the opinions enunciated by the 'savants' of that age, we think ourselves justified in coming to the conclusion that they did not know the true system, notwithstanding the laudable efforts made by Bailly to prove the contrary; assigning the knowledge the Greeks had to instruction they received from a highly civilized but, alas, extinct race called the Atlantides. This last part of his hypothesis renders refutation unnecessary. The following example will serve to illustrate the erroneous ideas conceived by some interpreters of the ancients prejudiced in their favour. Pythagoras had, amongst others, the very remarkable doctrine that the sun and stars, in their motions, emitted sounds, which varied in intensity according to their distances from the earth. Dr. Gregory draws from this a proof that the Pythagoreans were acquainted with the Newtonian law of the decrease of gravity. He observes that these philosophers spoke allegorically when they asserted that Apollo touched the seven-stringed lyre, which he supposes to represent the sun and the seven planets, and to indicate that the former retained the latter by attractive forces in harmonic proportion, and because the tones obtained from chords of equal thickness are inversely proportioned to the squares of their lengths, he infers that the harmonic proportion alluded to is the inverse duplicate of the distances of the planets from the sun. For our own part, we are not disposed to believe that the doctrines of Pythagoras, or of any of his disciples, lay as deep as Dr. Gregory imagined.

CHAPTER I.

THE DISCOVERY OF THE EARTH'S MOTION.

EVERYONE, however limited his thirst for knowledge, must be eager to learn whether this orb, which it has pleased the Creator to give us as a temporary home, is endowed with motion or not; and those who have already decided this question in their minds, must feel gratified either to find sympathy with their creed, or additional proofs in support of their opinions. Moreover, the propagation of the doctrine of the earth's motion presents one of the most conspicuous and peculiar passages in the history of the human intellect; for it is the first instance where the thinking world freed itself from deep-rooted illusions in which the greatest philosophers were spellbound — it is the first grand example of proud scholars renouncing long-established doctrines, and accepting — nay, teaching — precisely the contrary of what all, laity and priesthood, had believed and preached through so many centuries. The system of Copernicus has not only laid a new foundation for astronomy, but it has emboldened investigators to doubt any dogmas in natural science, it having been proved that man lived 5,000 years in blind and obstinate error, and given to the world the practical lesson first followed by our great countrymen, Bacon and Newton, only to take that for granted which is the result of proper deduction from a sufficient number of facts and observations.

The system of the celestial motions that prevailed before Copernicus was both complicated and erroneous. Pythagoras may be said to have been the first who entertained the idea that the earth moved and that the sun was fixed, but unfortunately the arguments brought forward in support of this doctrine by his school of philosophy were not sufficiently strong to withstand the attacks made upon them by Aristotle, who did not maintain that the stars and planets moved individually round the earth, but that they were all fastened to the concave surface of a crystal sphere, which rotated, carrying them with it in its course. Wild and fanciful as this theory must appear to us, it nevertheless was universally adopted and developed by successive generations of astronomers, up to the time of Purbach, in the middle of the fifteenth century.

At this period the earth was supposed to be immovably fixed in the centre of the universe; immediately surrounding it were the atmospheres of air and fire, and beyond these the sun, moon, and planets were supposed to travel round the earth, each fixed to a separate orb, or heaven, of solid but transparent matter. The stars were considered fixed in an outer orb, beyond which were two crystalline spheres; and on the outside of all the 'primum mobile,' which sphere was supposed to revolve round the earth in twenty-four hours, and by its friction, exercised on the interior orbs, to carry them round with a similar motion: hence the diversity of day and night. But, besides this principal and general motion, each orb had one of its own, which was intended to account for the apparent changes of position of the planets. This supposition, however, proving insufficient to account for all the irregularities, other hypotheses were introduced :—

Firstly. That to each planet belonged several concentric spheres.

Secondly. That the centres of these revolving spheres were placed on the circumference of a secondary revolving sphere, the centre of the latter being the earth.

Thus originated the names of Eccentrics and Epicycles.

The whole art of astronomers was now directed towards inventing and combining different eccentric and epicyclical motions, so as to represent the phenomena of the heavens. The name of Aristotle lent its powerful assistance to enable the false system to prevail against the true one, which was partially conceived by some philosophers before his time, as is gathered from his own writings.

Accordingly the heavens rapidly became, under this system, to use the words of Milton —

> With centric and eccentric scribbled o'er,
> Cycle and epicycle, orb in orb.

The bigotry of the middle ages accepted and protected willingly these heathen notions, rendering it most dangerous to confute them. Copernicus was well aware of this, and prudently delayed the publication of his immortal work, 'De Revolutionibus Orbium Cœlestium,' which appeared in 1543, the year of his death. His wish seems to have been to refute as gently as possible the opinions of the followers of Aristotle, who exercised an intellectual monopoly which he justly held to be absurd. It seems never to have occurred to him that an outcry might be raised against his system on religious grounds, for if so he would not have dedicated his book to Pope Paul III. or mentioned in the preface that he was induced to publish it by the persuasion of his friends the Cardinal of Capua and the Bishop of Culm. Apparently the prelates did not think that a case of impiety could be made out of an astronomical theory. But the Aristotelians were determined to punish as heresy that which they could not overturn by scientific reasons, and the clamour of irreligion which they set up deterred many astronomers from avowing openly his doctrines, though they were fully convinced of their sublime and simple truth.

In a letter from Galileo to Kepler, written from Padua in 1597, we find the following passage:—

'Congratulating you on your elegant discoveries, I shall only add a promise to peruse your book dispassionately; this I shall do the more willingly, because many years ago I became a convert to the opinions of Copernicus. I have arranged many arguments and confutations of the opposite opinions, which, however, I have not yet dared to publish, fearing the fate of our master Copernicus, who, although he has earned immortal fame among a few, yet by an infinite number (for so only can the number of fools be measured) is exploded and derided. If there were many such as you, I would venture to publish my speculations; but, since that is not so, I shall take time to consider it.'

This time for consideration lasted till about 1612, when the all-powerful love of truth vanquished his fears; and having observed some solar spots, he printed that discovery, and ventured in the same pamphlet to assert the justness of the Copernican system, adding some new arguments to confirm it. For this he was cited before the Inquisition, condemned to prison, but released after some months, on the promise that he would renounce his heretical opinions. However, in 1632, he published his 'Dialogues of the Two Systems of the World, Ptolemaic and Copernican.' He was again cited before the Inquisition, and sent to prison by that abominable court at Rome. In June the same year the Congregation convened and pronounced sentence against him and his books, obliging him to abjure what they termed his errors.

The following is the form of abjuration taken from Delambres' 'Histoire de l'Astronomie:' —

'I Galileo Galilei, son of the late Vincent Galileo of Florence, seventy years of age, being of sound mind and on my knees before you, most eminent and reverend Cardinals of the Universal Christian Republic, Inquisitors General of Heretic Malice, having before my eyes the Saints, and the Holy Testament in my hand, do swear that I always have believed, that I do now believe, and, with the grace of God, always shall believe, everything that is taught by the Holy Roman Catholic Church; but because this holy institution justly ordered me to abandon entirely the false opinion that the sun is the centre of the world and immovable, that the earth is not the centre and that it moves, and because I could neither support it nor defend it, nor teach it in any manner whatsoever, either by word of mouth or in writing, and after the said doctrine had been declared contrary to Holy Writ, I wrote and caused to be printed a book in which I treat of the said condemned doctrine, and give powerful reasons in favour of the said doctrine without bringing it to any result, it is for this that I have been strongly suspected of heresy, for asserting and believing that the sun was the centre of the world and immovable, and that the earth was not the centre and did move. It is for this reason, that wishing sincerely and from my heart to obliterate from the minds of your Eminences and all good Catholics the strong suspicion justly conceived against me, I do abjure, curse, and detest the errors and heresies aforesaid, and generally any other error or sect whatsoever contrary to the Holy Church aforesaid, and I do swear that in future neither by word of mouth or in writing will I state or affirm anything which could raise similar suspicions against me; and if I knew of any heretic or any one suspected of heresy, I would denounce him to the Holy Office, or to the Inquisitor, or to the priest of the place in which I might be. I swear, moreover, and promise that I will strictly fulfil and observe all penances which are or may be enjoined me by the Holy Office; if I should act contrary to any of these my words, promises, protestations, and oaths, which God forbid, I will submit to all pains and penalties which by the holy canons and other constitutions, general as well as special, have been instituted and used against such delinquents. So help me God and the Holy Apostles.

'At the Convent de Minerve, on the 22nd of June, 1633.

'(Signed) GALILEO GALILEI.'

Thus did a few contemptible bigots exercise their tyrannical authority over this venerable man (he was then seventy years of age), making him abjure doctrines which he knew to be founded on sound reason. Happily, however, for science, this infamous court has long been disarmed of its terrors, the Copernican system has prevailed, and the philosopher can now investigate the works of Nature, and benefit mankind by the expression of his opinions, unfettered alike by the cruel intolerance of Jesuitical tribunals like this, and the calumniating fanaticism of a bigoted priesthood. Since the time when the Reformation dispelled the clouds of superstition, science has advanced without requiring martyrs like Galileo to maintain that the earth rotates.

In the year 1737, a century after this monstrous sentence had been carried into effect, a more civilized generation rendered that justice to Galileo, dead, which their forefathers had refused him living, by erecting in the most conspicuous part of the church of Santa Croce a monument to perpetuate the memory of that great and persecuted philosopher. The erroneous notions of the ancients, so obstinately supported by the absurd superstitions of the middle ages, have vanished before the light of sound philosophy; but not without a struggle, for many direct arguments were brought to bear against the Copernican system.

One of the most plausible objections to the earth's motion seems to be that raised on the ground that we are unconscious of that motion. We can, however, easily demonstrate, by observation and experiment, that we may actually be moving without being aware of it, by attributing the change of the relative position of surrounding objects to a motion in themselves which they do not possess. For instance, a person travelling along a rough road is only reminded of the motion of the carriage by the roughness of the road, whereas in a vessel on a smooth sea he is led to believe, not that he is progressing, but that fixed objects on land are receding from him —

Provehimur portu terræque urbesque recedunt.

Even motion produced by our own exertions, such as in walking, is sometimes delusive. Strange as this may appear, the following experiment will suffice to illustrate it. Let a person walk close to railings, nearly his own height, so that their outline may be about on a level with his eye, then, as he moves along, he will see them continually dancing up and down, and the impression on his senses is that the railings possess in themselves a vertical motion. A little reflection will teach him that, by his mode of stepping, the eye is at different elevations at different periods, causing the apparent variation of level in the outline of the railings. It is easy to multiply such experiments, the results of which will be similar. In fact, we may come to the conclusion, that we are insensible of motion unless made aware of it by mechanical obstructions or muscular exertions.

It is obvious that when we treat of the question of the earth's motion, no evidence can be deduced from our own personal movements. Nor is it possible to show that mechanical impediments exist in the perfectly smooth frictionless path of our planet. Therefore we are led to believe that, whatever be the motion of the earth, we must be unconscious of it. Many other objections have been raised against the Copernican hypothesis, which at one time possessed weight and importance in proportion to the celebrity of their propounders. Riccioli collected seventy-seven arguments against the earth's motion, of which he himself refutes forty-nine as favourable to the Copernican system, and of all the remaining only those deserve any consideration which are founded on the argument of Ptolemy ('Almagest,' I. 7), viz., that the birds in the air would see the earth moving under them, and therefore never leave its surface, for fear of losing their nest and young; or, as Buchanan expresses it:—

> Ipsæ etiam volucres tranantes aera leni
> Remigio alarum; celeri vertigine terræ
> Abreptas gemerent sylvas, nidosque tenella
> Cum sobole, et cara forsan cum conjuge; nec se
> Auderet zephiro solus committere turtur.—*Sphæra*, i.

Tycho Brahe's principal objection is similar, though brought forward in another form. In a letter to Rothmann, he asked him, 'how it was possible that a ball dropped from the summit of a tower should always fall close to the foot of it, for if the earth completed an entire revolution in twenty-four hours, the tower must have advanced towards the east and moved a considerable distance from the ball in proportion to the time of descent of the latter.' Thus assuming the rate of the earth's motion at the equator to be eighteen miles per minute, as it is nearly, and the height of a tower to be 400 feet, which a body would take about five seconds to fall through, then the falling body would reach the ground 1½ miles westward from the foot of the tower, which is certainly contrary to all observation.

The fallacy of this reasoning is obvious, and it is astonishing that men of science should have fallen into an error like this, which could only have arisen from ignorance of the first principles of mechanics. It is a known fact, that when a stone drops from the mast of a vessel in motion, it falls at the foot of it, just as if the ship were at rest—and why? Because the motion of the ship was communicated to the stone before falling; and thus (supposing it to be possible) though birds might fly even higher than the atmosphere, they have received before that the impulse of the earth's motion, its direction and velocity, and must therefore continue to rotate with it.

The actual motion of a ball, while falling from any height, is compounded of two motions, one in the direction of the earth's rotation, the other in that of gravity; the path described, therefore, is a curved line. But as the change in the relative position only of the falling body to the earth is produced by the action of gravity, this result is the only one that can be observed; and the apparent motion is the perpendicular.

Another objection was raised, which at the present time appears childish, yet it is worthy of mention, inasmuch as it hindered many from becoming converts to the true opinions of Copernicus. The diurnal motion of the earth was imagined to be an absolute impossibility, for, reasoned the astronomers of former ages, 'if it were not so, after a lapse of twelve hours we should find ourselves head downwards'—a contingency which they contemplated with feelings of the deepest awe. We may, however, excuse their limited notions, for the voyages of discovery had not then informed them of the existence of antipodes, and that there is no difficulty in conceiving that we can rest in the same position as the New Zealanders. In the letter of Tycho Brahe, before alluded to, we find another objection to the earth's rotation founded upon the parallelism of its axis. 'How can it be imagined,' he writes, 'that one and the same body should have two motions so different, one which translates its centre of gravity, the other which changes the position of its axis?' The answer is, that the parallelism of the axis does not require a specific motion, as he supposed, for it is a fixed position.

It is sufficient that the axis has been once directed to a certain point of the heavens in order to remain in that direction, though the earth has an annual motion. There exists no physical or mathematical reason whence we might conclude that the axis of diurnal motion must be directed perpendicularly to the annual orbit, for between these two movements there is neither connection nor dependence. As soon as the particles composing a globe are projected simultaneously in the same path, they will acquire velocities and directions equal and parallel to each other, and hence do not undergo any change in their relative positions. Let us apply this reasoning to our earth; moving on a fixed axis, and hurled into space, all particles of it received the same impulse, and preserved their original rotation. This may be illustrated by a top spinning on a table; whatever motion is imparted to the latter, the top will continue to revolve on the same axis. So a ball projected from a cannon always preserves its axis of revolution. Having disposed of the objections which were raised against the Copernican doctrine on mechanical grounds, let us consider those which were founded on the testimony of Holy Writ, and which the fanaticism of the middle ages rendered perilous to refute.

The principal text for controversy was taken from the Book of Joshua, x. 13: 'And the sun stood still, and the moon stayed, until the people had avenged themselves upon their enemies. Is not this written in the book of Jasher? So the sun stood still in the midst of heaven, and hasted not to go down about a whole day.' Reading this verse, together with the one preceding it, in which it is stated that Joshua said before Israel, 'Sun, stand thou still,' &c., we find that he could not have expressed himself differently without making himself unintelligible to those he addressed.

Astronomers of the present day say 'the sun rises,' 'the sun sets,' without being reproached with misunderstanding the true state of nature. Joshua, conversing with his nation, could not speak otherwise than in their own language. It is absurd to maintain that a general — for such was Joshua — could give lessons of astronomy to his army, at the very moment when he was to show them the power and glory of the Supreme Being manifested in their favour; for if he had commanded the earth to stand still, he would have been obliged to explain the reason of his strange phraseology, so contrary to all usage. Therefore, if Joshua as a prophet knew that which in his time, and especially in his country, was unknown, he could not have expressed himself otherwise than he did.

Some other passages occur, such as in Isaiah xxxviii. 8, Ecclesiastes i. 5, which may appear contrary to the true hypothesis of the earth's motion, if taken in their literal sense and not in that in which they should be understood, namely, of ordinary parlance, or of poetical metaphor.

The number of texts which touch generally on the subjects of astronomy and natural philosophy is very large, but for the most part they are not scientific, but popular, suited to the common apprehension of mankind; for instance, David says, Psalm xviii., 'Tellus fundata super maria.' Solomon asserts, 'Terra in æternum *stat*;' lastly, in Kings xxiii. 7, we find it stated that a bath was constructed in a circular form, ten cubits in diameter, but measuring only thirty cubits in circumference, whereas mathematically, the latter must have been more than thirty-one cubits.

The deduction which we must draw from all these quotations is, that they will not bear scientific interpretation, as they were not meant to convey any scientific instruction; as when we read that the sun moved, it is obvious to us that the inspired writers did not aim at deciding a question of philosophy, nor did they intend to establish or ordain any opinion about it. These passages are either inferior in importance, or are not matters of faith, but are merely remarks accompanying an historical narration. After due and impartial consideration of the most celebrated objections brought during ages to bear against the theory of the earth's rotation, we must arrive at the conclusion that none of them, either moral or physical, are sufficiently cogent to impress our minds with the faintest idea that they are founded on reason or tenable in their consequences.

CHAPTER II.

THE PROBABILITIES OF THE EARTH'S MOTION.

THE earth being a globular mass floating freely in space, without pedestal or material suspension to keep it in its place, it follows that if once this globe received an impulse, it would move for ever in a straight line, since there is no friction and no resisting medium.

Now, if we grant the possibility of a power existing capable of communicating this impulse, we can easily imagine that this globe should be in motion.

But we may carry our speculation on the probabilities of this case much further.

Fig. 1.

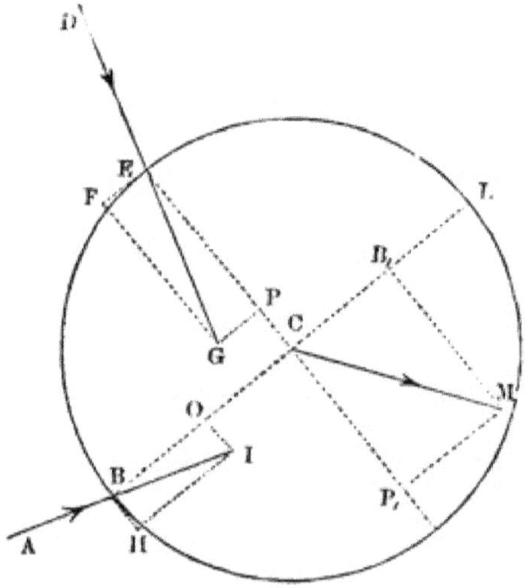

It is a principle of mechanics, that if motion be communicated to a mass in any other direction than through its centre of gravity, it will acquire two motions, one of translation, of which all the particles partake equally, the other of rotation, which affects only those which lie without the axis. It is a remarkable fact that these two motions are independent of each other, and if the different impulses of a central force were to act on a globe possessing these two motions, the effect would be only to change its rectilinear path of translation into a curvilinear orbit without disturbing the axis and velocity of rotation.

The mathematical investigations of these mechanical truths may be effected in the following elementary manner: —

Let E B L be a great circle of a homogeneous sphere, placed 'in vacuo,' and suppose the latter to receive an impulse represented by the line A B in magnitude and direction. It is plain that if A B were perpendicular to the surface of the globe, and consequently passing through its centre, the action of the force would make the globe move uniformly in that perpendicular; but in the case under consideration we assume the force to act obliquely, whence it follows that the force A B must be resolved into two, the one, B O, perpendicular to the surface of the sphere, the other, B H, parallel, i. e. tangential to it, so that the straight lines B O, B H become sides of a parallelogram H B O I, the diagonal of which is B I = A B. Now B O expresses the component of the force A B which impels the globe uniformly in the direction B C B_{1}, within the plane of the great circle B E L, termed the equator, whilst B H expresses the component of the force A B which gives to the point B at the surface of the globe a tendency to recede from the centre C along the tangent B H. Hence there is a force pushing the point B uniformly towards H, but at the same time this point B will, on account of the reaction of cohesion, return the same distance towards the centre C as the component B H made it depart from it. The result necessarily is that the point B must move in a circle uniformly about the point C, or, which is the same, about the axis passing through C. The oblique force therefore imparts to the globe two uniform motions, one which causes it to turn on its axis perpendicularly to the plane of the equator, termed 'rotation,' the other which causes it to advance in space in the direction of the plane of the equator, called 'translation.'

If the globe receive another impulse represented by the line D E in the plane of the same equator, this new force must also be resolved into two, the one, E P, pushing the globe in the direction E C with a velocity expressed by E P, the other, E F, tending to make the globe turn uniformly on the axis of that equator. Now making $CP_{,} = EP$ and $CB_{,} = OB$, then the two uniform forces of translation which act upon the globe are expressed by $CB_{,}$ and $CP_{,}$. Therefore the latter will move through the diagonal C M, which is in the plane of the equator, turning at the same time on the axis of it with a velocity expressed by $EF + BH$, since these forces together tend to the same effect and hold the same position relatively to the globe's surface. It is also obvious that the different degrees of obliquity of the impulse will produce different ratios in the velocity of rotation to that of translation; but any *central* force acting on a globe which is already endowed with the two motions of rotation and translation, will only affect the latter, as may be shown in the following way.

If a globe, the section of which is A P B Q, while revolving about the axis P Q, be impelled by a force in any direction C S, passing through its centre, and, consequently, perpendicular to its surface, it is evident that no part of this force can act tangentially to it, but its whole effect on the two hemispheres, A P B and A Q B, on which it acts equally, is only to direct them towards the point S

Fig. 2.

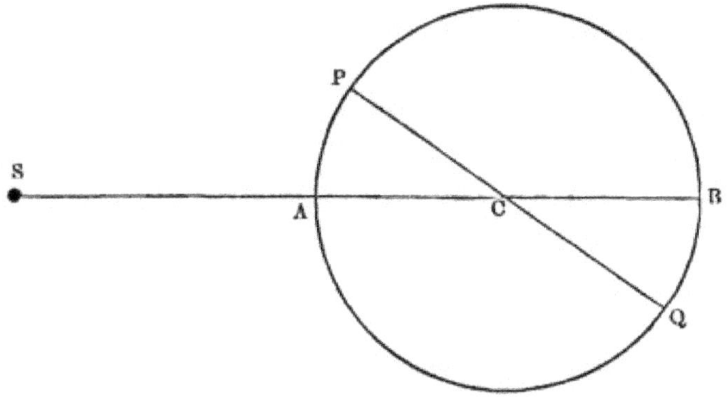

or away from it, according as the force is one of attraction or repulsion. Therefore this force must necessarily accelerate or retard the motion of translation, according as it acts with it in the same or opposite direction. Should, however, the line of action of the force at S include an angle with the rectilinear path of translation, then the direction of the latter will continually change, as is obvious from the theory of the composition of motions, and the globe in this case will describe a curvilinear orbit without modifying the motion of rotation.

Hence we can understand how the annual motion of our earth would not interfere with the diurnal motion, assuming that there be one. Let us now proceed to weigh the probabilities of the truth of this assumption. For the sake of illustration, let us imagine the force of gravity to be destroyed, and let us then suppose that a person holding a ball should open his hand and set it free. The result would be that it would commence spinning about one of its diameters, at the same time advancing uniformly through space. If the ball had received no impulse when the hand was opened, there being no force of gravity to attract it, it would have remained stationary in space. But we know from experience that it is impossible to open the hand without communicating some impulse, however slight, to the body grasped, and the chances against the direction of this impulse passing through the centre are infinite. We may therefore conclude, with a degree of probability almost amounting to certainty, that the result will be two motions — translation and rotation. Let us now apply this illustration to our earth. No one can deny that the Almighty hand which created it could have left it quiescent in space; but that it moves is by far the more rational hypothesis. For after the Creator had covered the face of this earth with animal and vegetable life, the alternations of day and night and the variations of the seasons became absolutely necessary for the preservation of that life, and since all these changes could be effected by a compound motion of rotation and translation resulting from an original impulse, and as all objections to such motion were shown in the preceding chapter to be groundless, it is on the whole more probable that it moves than that it is at rest. This probability is further increased by analogy. We have ocular proof of the rotation of other celestial bodies, by observing through telescopes the movement of spots on their surface. The moon turns upon its axis in about $27\frac{3}{10}$ days, the sun in $25\frac{1}{2}$ days, Mercury, Venus, and Mars in nearly a day, Jupiter in about $9\frac{9}{10}$ hours, and finally, Saturn in about $10\frac{1}{3}$ hours.* Moreover, all the revolutions of the planets about their axes take place in the same direction, from west to east, and our earth must rotate in a similar direction in order to produce the phenomena of day and night.

Having shown the last hypothesis to be highly probable, let us now consider the pros and cons of the other.

Either the globe revolves upon an axis every twenty-four hours, or the whole universe, including the sun, moon, planets, comets, and fixed stars, accomplishes a revolution around the earth in the same time: no third opinion upon the subject can possibly be held. In the latter case, it is evident, from the distance of the celestial bodies, that their diurnal revolution round our planet must involve a rate of motion utterly inconceivable. The sun must travel at the rate of 400,000 miles a minute, the nearer stars with a velocity of a thousand million miles in a second, and the more distant with a rapidity which no words can express.

It is, indeed, absurd to suppose this, when the end which was to be gained required only our little globe to revolve upon itself.

Is it probable that the planets, endowed with their own motion, the comets, which seem to have no resemblance with other celestial bodies, the millions of fixed stars dispersed throughout the firmament; all these, I say, which have no common relations with each other, which are placed at distances our imagination has difficulty to

* To the rapid rotation of these planets is attributable the great flattening at their poles, which may be distiuctly seen through the telescope, amounting, in Jupiter, to $\frac{1}{14}$ of its radius, and in Saturn to $\frac{1}{11}$.

conceive — is it probable that these should unite in one single mass, and perform a daily revolution round the atom which in the infinity of space represents our earth? No! The regularity and sameness in the motion of numberless bodies so different from each other in all other respects, indicate that these motions are not real; and the more we reflect upon it, the stronger becomes our conviction that the earth rotates.

This uniformity of motion, which appears to us so incomprehensible, was explained by the ancients by the assumption that the stars were fixed to hollow spheres made of a crystalline substance, and moving together one within the other; but since we have observed that the planets vary their distances from the earth, that at one period they advance towards it, while at another they recede from it, and that the comets sometimes come close to us and then disappear altogether, we may conclude that the hypothesis of a solid heaven is an absurdity. If, however, the heavens consist of millions of detached bodies, then the uniformity of their movements can be attributed to no physical cause, and the laws of motion which nature observes in all other things must in this case be checked by direct spiritual influence. We cannot hold with Riccioli, who, in his 'Almagest Nov.' (II. 248), states that the celestial bodies are carried round daily by intelligences called angels, specially employed in that behalf. Rejecting such mystical powers as incompatible with the dignity of true science, the following mechanical considerations present themselves. If the diurnal movement which we see be attributable to a motion of the stars, then all of them, except those lying in the equinoctial, will rotate in circles, the centres of which are situated without the earth — in fact, in an imaginary line termed the 'Celestial Axis;' and the same may be said with regard to the planets, since their orbits intersect the equinoctial in two points only. But whenever a body describes a circle with a uniform velocity, it is attracted towards the centre of its orbit by a constant force, the magnitude of which depends jointly upon its rate of motion and the radius of the circle it describes. The theorem relating to central forces was demonstrated by Huyghens, and is as follows, viz., that the centrifugal force which is equal and opposite to the centripetal, is directly proportional to the square of the velocity, and inversely proportional to the radius, or, in other words, it is inversely proportional to the square of the periodic time, and directly proportional to the radius. Since the periodic times of apparent rotation are the same for all the stars, namely, twenty-four hours, it follows that the magnitude of the force towards which each star is attracted will merely depend on the radius of the circle of declination which the latter describes. The central forces must therefore continually decrease, the nearer the stars upon which they act are to the poles of the heavens, and the locus of all these different forces extending from both sides of our earth to infinity is nothing but an imaginary line without physical dependence. It is against our notions of mechanics to suppose that a multitude of moving forces should have their seat and origin in as many mathematical points, and yet, if we admit the immobility of the earth, we must accept this hypothesis. From these and the preceding considerations, it follows that an actual diurnal rotation of the firmament is highly improbable, and this improbability, great in itself, is infinitely strengthened by the probability of the other hypothesis, that the earth rotates. But conclusive as may be this balance of probabilities, the question admits of more rigid investigation, for there exist direct, positive, and even experimental demonstrations of the earth's rotation about its axis, the elucidation of which will form the subject of the following chapters.

CHAPTER III.

PROOF OF THE EARTH'S ROTATION BY THE TRANSMISSION OF LIGHT.

WE have stated in the last chapter that the apparent phenomena of the heavens must be the same whether the latter complete an entire revolution from east to west in twenty-four hours, the earth remaining fixed, or whether the earth itself performs an entire revolution in the same time. This proposition is true, if the transmission of light be instantaneous, however great the distance through which it has to pass. But if the velocity of light be not infinite, that is to say, if the rays take some perceptible time to travel from the stars to us, then this proposition does not hold good, as Arago has proved by the following ingenious method of reasoning. Let us suppose our earth to be fixed, and a star to move round it from east to west. The star is always the centre of diverging rays, but the position of this centre relatively to the horizon and the meridian of a given place will vary continually. The rays emanating from the star move in straight lines, and the star itself will therefore become visible on the horizon by those rays of light which were emitted when it actually was in the horizon; in the same manner it will appear in the meridian, by rays lying exactly within the plane of the meridian, and this will occur at the moment of its real culmination. We may deduce from this, that if the earth does not rotate, but the heavens do, a star can only be seen in the horizon or meridian by means of those rays which it emitted when actually in the prolongation of either of those planes. Let us now suppose the velocity of light to be such, that for a given star it requires six hours to reach us, then its culmination would be apparent to us six hours after its actual transit; therefore, if a star were at such a distance that its light required twelve hours to reach us, it would appear to rise when in reality it was setting; and if its distance were still greater, it would be seen in the east long after it had passed below the horizon. Following up this reasoning, and assuming the distance of two stars from our earth to be different and proportional to their real positions, then, though to us they may appear to be side by side, yet in reality they may be separated by distances scarcely expressible, even by taking the diameter of our orbit as the unit of measure.

Before examining whether these results derived from the double assumption of the fixity of the earth and the successive propagation of light are reconcileable with the facts observed, let us consider the case in which the earth moves and the firmament is fixed. The centres whence the light proceeds in straight lines are fixed in space, therefore any of these luminous points will appear to rise when the horizon in its motion of rotation from west to east coincides with any of the descending rays; and similarly a star will pass the meridian when the prolongation of this plane coincides with the invariable position towards which all the rays, which render it visible to us, converge. With regard to the rise of a star and its culmination, it matters not whether the luminous particles, by which these phenomena are produced, quitted the star days, years, or even centuries before they were observed, since they travel in straight lines terminating in fixed points of the firmament in which the stars are situated. Hence the velocity of light does not, in the latter case, exercise any influence on their apparent positions. When a star appears to pass the meridian, it really does so, and when two stars seem to be near each other, the lines drawn to their centres from the earth are really in close proximity.

The deductions arising from the hypothesis of the fixity of the earth must have appeared strange even to those who made them, but in scientific as in other researches, the singularity of a proposition is no proof of its fallacy. Let us now consider if we cannot trace some facts in connection with the motion of the stars, which are irreconcileable with the deduction that their apparent position must depend upon their distance from the earth; for instance, the apparent time of culmination of Mars, when in 'conjunction,' is equal to the time of its real transit plus the time which light takes to pass from it to the earth, and the same is true when the planet is in 'opposition.' But the distance of Mars from the earth at the time of conjunction is less than that at the time of its opposition by double the distance of the sun from the earth. Therefore, comparing the observed with the real transit, we should have an inequality between the 'opposition' and the 'conjunction' amounting to about 16·5 minutes of time, that is, double the time required for light to travel from the sun to the earth. It is further evident that from the same reason the apparent motion of the planet between its conjunction and opposition must take place from east to west, but the existence of such inequalities is in no way indicated by observation; similar perturbations must take place with Jupiter and Saturn, and, with regard to the double stars, it may easily be shown that the hypothesis of the earth's immobility would lead to results still more inadmissible. When the principal star and its satellite are both at the same distance from the earth, they will appear very close together (which they in reality are), but if the satellite in the course of its revolution round the principal star takes a position by which its distance from the earth is increased by a quantity equal to the diameter of the terrestrial orbit, then, for seeing it almost in contact with the centre of its motion, it ought to appear distant in right ascension by a quantity which, expressed in time, would amount to more than eight minutes,—a result so discordant with all observed facts, that we may consider it as a mathematical demonstration of the falsity of the hypothesis from which we started, namely, the immobility of the earth.

An objection might be raised to this mode of reasoning, namely, that the determination of the velocity of light presupposes a knowledge of the true system of the world, and it would have been a very important one in the days of Römer, when the measure of that velocity was solely founded upon the observations of the eclipses of Jupiter's satellites. But now that recent researches have shown us how to ascertain the velocity of light on our own globe, this new method, first suggested by Arago, seems to rest on a firm basis.

This great advocate of the undulatory theory considered the retardation of light in dense media a crucial proof between the rival hypothesis of Newton and Huyghens. He, therefore, as early as 1838, recommended the application of Mr. Wheatstone's admirable invention of the revolving mirror as a means of measuring extremely small intervals of time, in order to compare the velocity of light in air with that in a corresponding length of water.

Professor Wheatstone contrived and used the apparatus alluded to in the year 1834, for measuring the velocity of electrical conduction. The principle upon which it was constructed may be understood from the following description. Let a copper wire, half a mile long, be so convoluted that the middle and the two ends may be brought near together, the whole being perfectly insulated. Let the wire be interrupted at these three places, which are so arranged as to lie in one line, and then be placed into connection with an electric machine or battery. The contact will effect three sparks, which, taking place close to each other, can easily be seen at once reflected in a small plane mirror. Let now the latter be put in very rapid rotation round a horizontal axis, so placed that the sparks, if possible, may be reflected together to the eye of the observer. In Mr. Wheatstone's apparatus the velocity reached 800 times in a second, consequently the mirror described one degree in $\frac{1}{800 \times 360} = \frac{1}{288000}$th part of a second. But for one degree of rotation of the mirror the reflected image will describe an arc of two degrees. If then the three sparks occur at the same instant of time, they will be seen in one line, but if either spark occur later than the others by only $\frac{1}{288000}$th of a second, the mirror will have revolved so much in this interval as to displace the image of that spark relatively to the others, by the very palpable angular amount of two degrees. In the copper wire half a mile long the end-sparks took place simultaneously, while the middle spark occurred later by about one-millionth part of a second, giving a velocity of transmission somewhat greater than that of light.

It will be understood, from the account of this method, that the retardation of light is shown by the displacement of the image of an object seen through water relatively to the image of the same object seen through air. Suppose the mirror to be turned from left to right, then, if light moves faster in water, as Newton thought, the water-image will be to the right, but if slower, according to Huyghen's theory, it will be to the left. The amount of angular deviation to be appreciated may be increased with the same speed of rotation, by applying two or more rotating mirrors. Regard, of course, must be taken to the loss of light by reflection and refraction. According to the experiments of Bouguer, light passing through ten feet of water preserves only $\frac{3}{5}$ths of its original intensity. When, therefore, Arago proposed the application of Mr. Wheatstone's revolving apparatus for a purpose different from that for which the inventor used it, he endeavoured first to prove the possibility of making sensible the slight differences of angular deviation, without surpassing the limits of the velocity of rotation and those of the length of the tube of water. His calculations on the subject are contained in a memoir presented by him to the Académie des Sciences, on December 3rd, 1838. The results of these show that one mirror, revolving a thousand times per second, requires a tube of water ninety-two feet long, in order to cause a deviation of one minute; but this length of the tube may be easily reduced to its half, third, or fourth, &c., by having recourse respectively to two, three, or four mirrors, &c., which must be done in all those cases where the image appears too faint. The practicability of the plan was therefore sufficiently proved by its projector; but perhaps, owing to his failing sight, he never realised his own idea. It was adopted by M. Foucault, who succeeded, in 1850, in demonstrating the retardation of light in a tube of water only 6½ feet long, and with a velocity of rotation not exceeding 200 turns. The latter, however, was increased for determining the absolute rate of its transmission.

It will not be out of place here to explain M. Foucault's combination of fixed and revolving mirrors, by means of which he was enabled to ascertain the prodigious rapidity with which light travels. The construction of the essential portions of M. Foucault's apparatus will be understood by the annexed figure (3). The camera obscura, C, has a

Fig. 3.

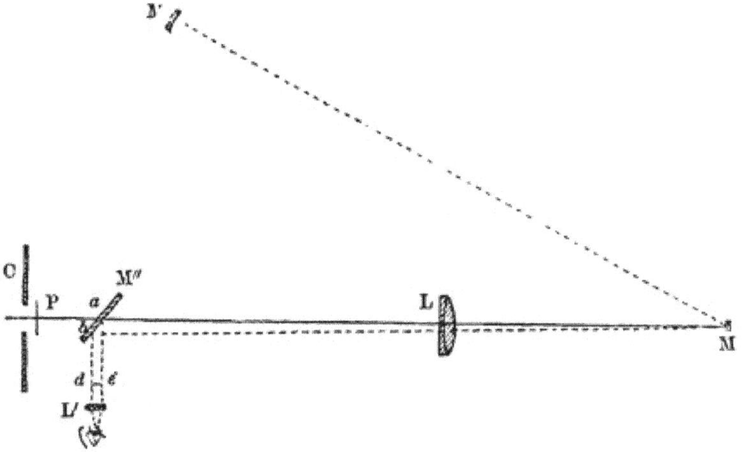

FOUCAULT'S ARRANGEMENT FOR MEASURING THE VELOCITY OF LIGHT.

small square opening, behind which is suspended a fine platinum wire P; a pencil of solar rays, reflected from a mirror placed outside, passes through the square opening into the camera, and, meeting the platinum wire, is thrown on to an achromatic lens L, placed at a distance from the wire less than double its focal distance; then the image of the wire, more or less magnified, will tend to form itself on the axis of the lens, but the pencil of rays, after having passed through the glass, meets a plane mirror M, revolving with great velocity, from which it is reflected, and forms an image of the wire in space which moves with an angular velocity equal to double that of the mirror. This is obvious from the following geometrical consideration.

Let M M (*fig.* 4) be the revolving mirror, P a fixed object forming its image in I, P I being perpendicular to M M, then

FIG. 4.

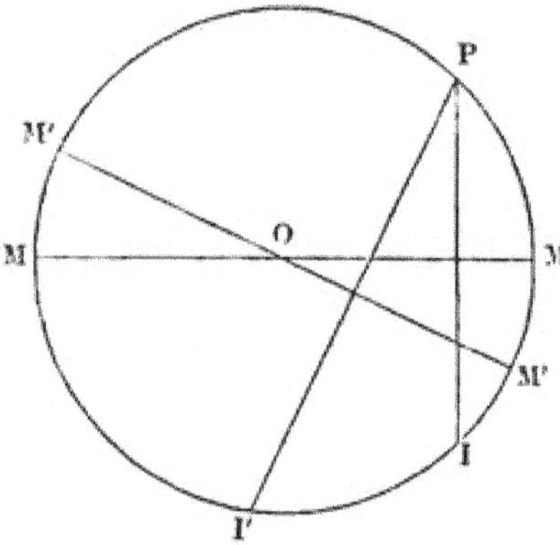

the mirror having come into the position M'M', the image will be formed at I', P I' being perpendicular to M'M'. Now ∠ M O M' = ∠ I P I', because the sides of the one are perpendicular to the sides of the other; hence, by Euclid, Lib. 3, Prop. 20., the arc I I' is double the arc MM', that is, the image travels with an angular velocity double that of the mirror. (Q. E. D.)

The image is again reflected in space (see *fig.* 3) by M', a fixed concave mirror, of which the centre of curvature lies in the axis of rotation of the revolving mirror, and coincides with its centre. The pencil of rays reflected from the mirror M′ turns back on itself, is again reflected from M, passes the lens a second time, and forms an image of the wire upon the wire itself as long as the mirror M turns slowly. In order to observe this image without obscuring the pencil of rays which passes on the camera at C, a plane mirror M″ is placed between the platinum and the lens at such inclination, that the reflected rays fall upon a powerful eye-piece L′. Now, if the mirror M is at rest, or is only turned slowly, the returning ray M M′ will meet the mirror M in the same place as during the first reflection; it will therefore take its original direction, meeting the mirror M″ in *a*, whence it is partially reflected, and forms in *c*, at a distance $a c = a$ P, the image which the eye observes at L′. The image appears at every revolution of M, and when its velocity of rotation is uniform, remains fixed in space. For velocities which do not surpass thirty revolutions per second, the successive phenomena are distinct, but above thirty the impressions on the retina are continuous, and give the body the appearance of being absolutely at rest. Finally, if the mirror turns with sufficient rapidity, its position is sensibly changed during the time the light required to travel from M to M and back again; the return ray, after its reflection from the mirror M, takes the direction M *b*, and forms its image in *d*, hence the image has undergone a total deviation *c d*. Strictly speaking, a deviation ought to occur as soon as the mirror turns, however slowly, but it will only be appreciable when it has attained a certain magnitude, which requires either the rotation to be rapid, or the distance M M′ to be sufficiently large. In Foucault's experiment, the distance between the two mirrors was

13 feet, and by communicating to the mirror M a velocity of between 600 to 800 revolutions per second, the deviations obtained amounted to about 0·01 inches. Denoting the velocity of light by V, the deviation by δ, and the number of revolutions per second by n, and assuming M M$' = l''$, M L $= l$ and L P $= r$, Foucault arrived at the formula

$$V = \frac{8\pi n r l^2}{(l+l')\delta};$$

by means of which be obtained results respecting the velocity of light closely approaching those derived from observations of the eclipses of Jupiter's satellites. The successive propagation of light being thus established without astronomical observation, forms therefore a legitimate premise in our demonstration of the earth's rotation.

We are inclined to think that D'Alembert was the first who drew attention to the fact that if the velocity of light be finite, and the earth be at rest, the stars could not be seen in their real positions, though Montucla asserts that Empedocles was the author of this opinion; however, be this as it may, these remarks would have remained inconclusive and sterile, had not Arago singled out phenomena in the celestial motions as necessary consequences of the hypothesis of the earth's fixity totally at variance with observed facts. Bradley's discovery and ingenious explanation of the aberration of the stars afforded the first striking evidence of the translatory motion of the earth, and nothing can be more interesting, or mark more highly the progressive character of science, than the fact that the same hypothesis, namely, the finite velocity of light, which led Bradley to his result, should years afterwards serve to confirm, with equal strength, its rotatory motion.

CHAPTER IV.

THE EARTH'S ROTATION DEMONSTRATED BY THE VARYING FORCE OF GRAVITY IN DIFFERENT LATITUDES.

THE first direct proof of the earth's diurnal motion is connected with the investigation of its figure. Picard, in his measurement of the earth, published in 1671, speaks of a conjecture proposed to the Academy, 'that assuming the rotation of the earth to be true, heavy bodies should descend with less force at the equator than at the poles;' and observes, for the same reason, 'there should be a difference in the length of the pendulum vibrating seconds in different latitudes. In the same year Richer was sent by the Academy of Paris to Cayenne, on a scientific mission; and one of the several objects of his voyage was to ascertain the length of the pendulum vibrating seconds. He returned in 1672, and mentions the observations of the pendulum as the most important he made. The same measure, marked at Cayenne on a rod of iron, according to the length of the seconds' pendulum for that place, having been brought back and compared with that marked at Paris: the difference was found to be 1¼ lines, that at Cayenne being the shorter. The arcs of vibration of the pendulum with which the experiment was made were very small; they continued for fifty-two minutes, and were compared with an excellent clock. Moreover, the clock which Richer took to Cayenne, having been adjusted to beat seconds at Paris, retarded there two minutes a day; so that no doubt remained of the diminution of gravity at the equator.

Newton, whose sagacity at once recognised the cause of this phenomena, did not hesitate to declare it as the experimental demonstration of the earth's rotation, so long sought for by men of science.

It is evident that all places on or near the equator must revolve quicker than those situate near the poles; and hence the centrifugal force being increased by the revolution in greater circles, must proportionately neutralise the centripetal force; i. e. the gravity towards the centre of the earth, upon which the oscillations of the pendulum depend.

Richer's experiment has since been repeated by other scientific men with the greatest care, and has always yielded the same results. It must not be supposed that to swing a pendulum, count its vibrations, and measure its length is an easy task. Experience shows that a more difficult practical problem can hardly be proposed, and that persons therefore who have overcome these difficulties rank high as promoters of science; such are especially Lacaille, Borda and Biot in France, and Foster, Kater, Sabine, and Bayly in England.

The accuracy with which these observations must be made, and the extreme delicacy of the instruments required, will be seen by the following description of Biot's pendulum. It is constructed on Borda's principle, which consists in making it as near as possible resemble the simple pendulum. The bob is spherical, and made of platinum, suspended by a metallic wire, the lower end of which is fastened by means of a screw into the neck of the spherical segment. The upper end of the wire is attached to another screw, which forms part of the prismatic bar of suspension. The edge of the latter rests on an agate plane, to which the oscillations are perpendicular. This plane of suspension is placed upon a plate of iron, and can be exactly adjusted by a small spirit level and parallel screws, the plate itself being supported by standards. The prismatic suspension is adjusted so that the duration of oscillation corresponds nearly with that of a clock-pendulum; this is effected by turning a small nut, screwed on the top of the knife-edge. When the times of oscillation coincide approximately, the wire, with the bob attached, is suspended from the rod immediately under the prism by means of the screw. The length of the wire must be so regulated that the oscillations of the entire system differ very slightly from those of the clock, and consequently from those of the prismatic suspension.

As the centre of gravity of the knife-edge is extremely near to the plane of suspension, its mass can, under the conditions by which the apparatus acts, exercise no appreciable influence upon the length of the pendulum. At a distance of twenty-five feet from the instrument, a telescope is placed, the eye-glass of which is intersected in its focus by a vertical wire, which must be made to coincide with the thread of the pendulum when in a state of rest, at the same time a small piece of paper is fixed upon the bob of the clock-pendulum to serve as an indicator. These preparations being completed, the clock-work is set in motion, and the simple pendulum also. The telescope enables us to judge the difference between the oscillations of the clock-pendulum and those of the simple one; it also increases the apparent velocity of the thread of the pendulum relatively to that of the indicator of the bob, and finally, by its means, we can ascertain exactly the times of coincidence of the two threads. But it is evident that the pendulum must either lose or gain upon the clock two oscillations during an interval of two consecutive coincidences, and hence the entire number of oscillations lost or gained by the pendulum, during a mean solar day, can be computed by simple proportion. Since the time of the oscillations of the pendulum increases with the magnitude of the circular arcs described, for only the vibrations in cycloidal arcs are isochronous, it becomes necessary to observe the amplitude at the moment of each coincidence. For this purpose a small circular scale, divided into minutes, is fixed to the bob of the clock, by means of which the observer can easily ascertain the magnitude of the arcs described by the pendulum.

In order to measure the length of the pendulum from the plane of suspension down to the globe of platinum, the following apparatus is employed. A metallic plane, resting upon supports, is placed at the commencement of the experiment just below the pendulum; this can be raised or lowered by means of a screw, of which the number of turns and the fractional part of each turn are indicated by a divided circle, which moves in front of a vertical rule similarly divided. A few trials will in general be sufficient to obtain an exact contact between the metallic plane and the platinum ball, great care being taken that the latter be not raised; for if it be, the effect of the tension of the ball upon the thread would be destroyed, and it could only be restored to its original state of rest after a considerable time. To ascertain when the contact of the ball and the plane is complete, the eye must be so placed as to observe any ray of light between the ball and the plane; the moment such ray disappears, the contact is perfect. In order to measure the distance of this plane from that of suspension, a sheet-rule is employed, at one end of which a knife-edge suspension is adapted, at the other a metallic index, which can be drawn up and down by means of a screw; both the index and rule are graduated equally. The pendulum is now removed, the rule suspended in its place, and the index lowered until it comes in perfect contact with the plane underneath; then, reading off the number of degrees the index has moved, the distance required will be known; subtracting from this result the radius of the platinum ball, the length of the pendulum is obtained. English observers generally prefer an invariable pendulum, in the form of a flat bar, provided with a weight. A pendulum of this description, after being compared with an astronomical clock at Greenwich, or any other observatory, is taken to other stations, and the rate of its vibrations determined; it is then brought back to the first place, and again compared with the chronometer. This method requires the reduction of the length of the compound pendulum used to that of the corresponding simple pendulum—a process by no means easy, considering the great accuracy requisite. Captain Kater's ingenuity overcame this difficulty, by applying Huyghens' beautiful theorem of the convertibility of the centres of oscillation and suspension to practical use. The pendulum he employed is represented in the annexed diagram (*fig.* 5).

The bar *cd* is furnished with two axles, by either of

which it may be suspended, one passing through *c* and the other through *o*. Besides the principal weight, *d*, it is provided with a small sliding-weight, *f*, which can be moved along the stem *c d*, and this weight is to be moved till the number of oscillations in a given time (as twenty-four hours), is the same whether the pendulum is suspended from *c* or from *o*. If *f* be placed in such a position that, by moving it from *o* as to *f*, the number of oscillations about *c*, in twenty-four hours, will be increased, and by the same change the number of oscillations about *o*, in the same time, will be still more increased. The adjustment is thus made:—Let the weight be at *f*, and let the number of oscillations in ten minutes about *c* be 606, and about *o* be 601; now let the weight be moved to *f*, and let the oscillations in ten minutes be 607 about *c* and 609 about *o* (because the latter are more affected than the former); then the proper position of the slider is somewhere between the first and last points. Let it be placed at *f*, and let the oscillations in this case be 606½ and 606; then the proper position is somewhat changed, observing always that if the number of vibrations about *c* be greater the slider must move towards *c*; and, if the contrary, it must move towards *o*. By this means, continually halving the distance last moved, we may make the oscillations about *c* and *o* approach within any required degree of exactness; the distance between *c* and *o* being then measured will give the length of a pendulum which makes a known number of oscillations in ten minutes.

FIG. 5.

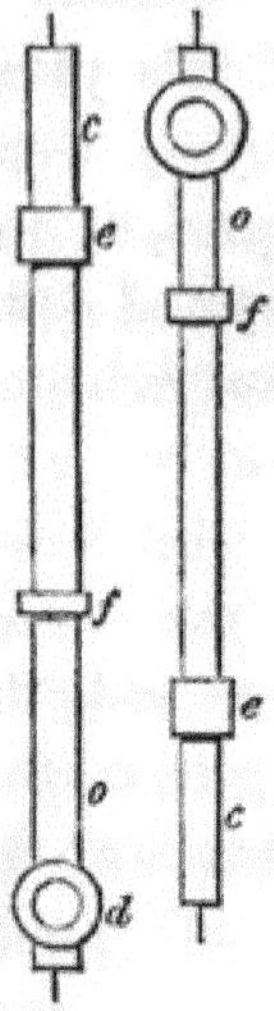

With a pendulum of this description, Captain Kater made his important experiments. In pursuance of the resolution of the House of Commons of March 15, 1816, the Royal Society of London requested this active astronomer to ascertain the length of the seconds' pendulum at Unst, Portsoy, Leith Fort, Clifton, Arbury Hill, and Shanklin Farm. He arrived at Leith on June 28, and reached Unst on July 9. After finishing his observations in the Orkneys, he arrived at Portsoy on August 1; at Edinburgh, August 28; at Clifton, September 28; at Arbury Hill, October 15; and at Shanklin Farm, in the Isle of Wight, about May 10, 1819, where he completed the series of laborious experiments which he had undertaken. The following are the results of these operations, as given by Captain Kater himself in the 'Philosophical Transactions' for 1819, Part III.:—

'It now remains to give in one view the results of the operations that have been detailed. These are comprised in the following table. It would have been desirable to have expressed the length of the pendulum vibrating seconds, in parts of the scale which forms the basis of the Trigonometrical Survey of Great Britain, the Commissioners of Weights and Measures having agreed to recommend that the "*standard* used in the Trigonometrical Survey of Great Britain should be considered as affording the most authentic determination of the linear measure of the United Kingdom." But, as experiments are yet wanting to enable me to do this with sufficient accuracy, I have given the length of the pendulum in parts of Sir George Shuckburgh's standard scale, the correction for the difference between which and the national standard of linear measure may be readily applied hereafter.

'The length of the pendulum vibrating seconds in the latitude of London is stated in the "Philosophical Transactions" for 1818, to be 39·13860 inches. But I have here to notice a very important omission, which I am obliged to Mr. Troughton for having pointed out in the first number of the "Edinburgh Philosophical Journal." It may be seen that in computing the specific gravity of the pendulum I have neglected to include the deal ends. Anxious to supply this omission in the most unexceptionable manner, I thought it best to take the specific gravity of the whole pendulum, and for this purpose requested Mr. Barton, comptroller of his Majesty's Mint, to allow me the use of the fine balance lately constructed under his directions, a request with which he most obligingly complied, and favoured me with his assistance, and with every requisite for making the experiment.

'A deal trough was prepared, seven feet long, nine inches wide, and the same depth. The pendulum was slung horizontally from the scale pan, by a fine iron wire. The weight of the whole was carefully determined in air, and found to be 66904 grains. The trough, which had been previously placed beneath the pendulum, was then filled with distilled water, and the weight of water displaced was found to be 9066 grains. The small portion of iron wire which was immersed in the water was carefully noted; the weight of the wire by which the pendulum was suspended was 56 grains, and the weight of water equal in bulk to that part of the wire which was immersed was 2·5 grains. The temperature of the water was 68°, and that of the atmosphere 62°; the barometer 29·9 inches. Hence we have the weight of the pendulum 66858·8 grains in vacuo, at the temperature of 62°; the weight of an equal bulk of water at the same temperature

9068·4 grains; and the resulting specific gravity of the pendulum 7·3727. Employing this specific gravity in computing the allowance for the mean buoyancy of the atmosphere, we obtain ·00624 for this correction instead of ·00545, the former erroneous conclusion. Besides this, the allowance + ·00031 for the height of the pendulum above the level of the sea should, according to Dr. Young's investigation, have been multiplied by $\frac{66}{100}$, making + ·00021 of an inch. These corrections being applied, we have 39·13929 inches of Sir G. Shuckburgh's standard scale, for the length of the pendulum vibrating seconds in the latitude of London.

'Wishing to compare with this the result which would have been obtained by means of the weights and specific gravities of the different parts of the pendulum, I carefully measured the deal ends, and found them to contain 3·956 cubic inches. The weight of the knife edges was 370 grains, and their specific gravity 7·84. With these data, and taking the specific gravity of deal at 0·49, the specific gravity of the whole pendulum will be found to vary from the more accurate determination above given a quantity which would have occasioned a difference in the length of the seconds' pendulum of only $\frac{1}{50000}$ of an inch.'

Place of Observation.	Latitude.	Vibrations in a mean solar day.	Length of the Pendulum vibrating seconds in parts of Sir George Shuckburgh's scale.
	° ′ ″		Inches.
Unst . . .	60 45 28·01	86096·90	39·17146
Portsoy . .	57 40 58·65	86086·05	39·16159
Leith Fort . .	55 58 40·80	86079·40	39·15554
Clifton . .	53 27 43·12	86068·90	39·14600
Arbury Hill .	52 12 55·32	86065·05	39·14250
London . .	51 31 8·40	86061·52	39·13929
Shanklin Farm .	50 37 23·94	86058·07	39·13614

Names of the Stations.		Diminution of Gravity from Pole to Equator.	Compression.
Unst and Portsoy		·0053639	$\frac{1}{304 \cdot 3}$
	Leith Fort	·0054840	$\frac{1}{315 \cdot 8}$
	Clifton	·0056340	$\frac{1}{331 \cdot 5}$
	Arbury Hill	·0054282	$\frac{1}{310 \cdot 3}$
	London	·0055510	$\frac{1}{322 \cdot 7}$
	Dunnose	·0055262	$\frac{1}{320 \cdot 1}$
Portsoy and Leith Fort		·0056920	$\frac{1}{338 \cdot 0}$
	Clifton	·0058194	$\frac{1}{353 \cdot 2}$
	Arbury Hill	·0054620	$\frac{1}{313 \cdot 7}$
	London	·0056382	$\frac{1}{332 \cdot 0}$
	Dunnose	·0055920	$\frac{1}{326 \cdot 9}$
Leith Fort and Clifton		·0059033	$\frac{1}{364 \cdot 0}$
	Arbury Hill	·0053615	$\frac{1}{304 \cdot 1}$
	London	·0056186	$\frac{1}{329 \cdot 8}$
	Dunnose	·0055614	$\frac{1}{323 \cdot 7}$
Clifton and Arbury Hill		·0042956	$\frac{1}{229 \cdot 6}$
	London	·0052590	$\frac{1}{294 \cdot 9}$
	Dunnose	·0052616	$\frac{1}{295 \cdot 1}$
Arbury Hill and London		·0069767	$\frac{1}{597 \cdot 5}$
	Dunnose	·0060212	$\frac{1}{380 \cdot 3}$
London and Dunnose		·0052837	$\frac{1}{297 \cdot 0}$

By comparing the observations of each station successively with those at all the others, Captain Kater obtained the results in the opposite table, containing the diminution of gravity and the resulting compression.

From the number of experiments on the length of the seconds' pendulum at various latitudes made for the English Government by Kater and Sabine, and for the French by Biot, Arago, Duperrey, and Freycinet, the following simple equation has been deduced. If L denote the length of a seconds' pendulum at any latitude λ, and 39,017 inches be the length of a seconds' pendulum at the equator, then

$$L = 39{,}017 + 0{,}200 \sin^2\lambda.$$

The numerical values of this formula have been derived from most accurate observations, the results of which are exhibited in the following table. According to Mr. Airy all the different lengths for the seconds' pendulum, there noted and arranged in the order of latitude, are, with one exception, erroneous to the amount of the error in the correction for the density of the air. The magnitude of this error is nearly the same for all, and is about 0,0018, except the length given by Bessel, which is accurate, owing to this astronomer's method of determining the atmospheric correction. It consists in observing the length of the seconds' pendulum in air, first with a bob of brass, and, secondly, with a bob of ivory. The correction, being greater for the latter than for the former in the proportion of the specific gravities, is then calculated with great accuracy from the difference of the observed lengths: —

TABLE OF THE LENGTH OF THE SECONDS' PENDULUM.

Place.	Latitude.	Length in Inches.	Observer.
1. Spitzbergen . .	79° 50′ N.	39·21469	Sabine.
2. Melville Island .	74 47	39·20700	Sabine.
3. Greenland . .	74 32	39·20335	Sabine.
4. Port Bowen . .	73 14	39·20419	Foster.
5. Hammerfest . .	70 40	39·19475	Sabine.
6. Drontheim . .	63 26	39·17456	Sabine.
7. Unst . . .	60 45	39·17162	Biot.
8 Stockholm . .	59 21	39·16541	Svanberg.
9. Portsoy . . .	57 41	39·16159	Kater.
10. Leith Fort . .	55 59	39·15546	Kater.
11. Königsberg . .	54 42	39·15072	Bessel.
12. Clifton . . .	53 28	39·14600	Kater.
13. Arbury Hill . .	52 13	39·14250	Kater.
14. London . . .	51 31	39·13929	Kater.
15. Greenwich . .	51 29	39·13983	Sabine.
16. Dunkirk . .	51 2	39·13773	Biot.
17. Isle of Wight .	50 37	39·13614	Kater.
18. Paris . . .	48 50	39·12864	Borda.
19. Clermont . .	45 47	39·11615	Biot.
20. Milan . . .	45 28	39·11603	Biot.
21. Padua . . .	45 24	39·11896	Biot.
22. Fiume . . .	45 19	39·11788	Biot.
23. Bordeaux . .	44 50	39·11296	Biot.
24. Figeac . . .	44 37	39·11215	Biot.
25. Toulon . . .	43 7	39·10952	Duperrey.
26. Barcelona . .	41 23	39·10432	Biot.
27. New York . .	40 43	39·10120	Sabine.
28. Formentera . .	38 40	39·09510	Biot.
29. Lipari . . .	38 29	39·09828	Biot.
30. California . .	21 30	39·03829	Hall.
31. Sandwich Islands .	20 52	39·04690	Freycinet.
32. Jamaica . .	17 56	39·03503	Sabine.
33. Marian Islands .	13 28	39·03379	Freycinet.
34. Madras . . .	13 4	39·02630	Goldingham.
35. Trinidad . .	10 39	39·01888	Sabine.
36. Sierra Leone . .	8 30	39·01997	Sabine.
37. Galapagas . .	0 32	39·01717	Hall.
38. St. Thomas . .	0 25	39·02074	Sabine.
39. Pulo Gaunsah Lout	0 2	39·02126	Goldingham.
40. Rawak . . .	0 2 S.	39·01433	Freycinet.
41. Maranham . .	2 32	39·01213	Sabine.
42. Ascension . .	7 55	39·02363	Sabine.
43. Bahia . . .	12 59	39·02433	Sabine.
44. Isle of France .	20 10	39·04684	Freycinet.
45. Rio de Janeiro .	22 55	39·04350	Freycinet.
46. Paramatta . .	33 49	39·07452	Rumker.
47. Port Jackson . .	33 52	39·07919	Duperrey.
48. Cape of Good Hope	33 55	39·07800	Freycinet.
49. Isles Malouines .	51 35	39·13781	Freycinet.

Since at the pole $\lambda=90°$, and therefore $\sin^2\lambda=1$, we have by substitution in the formula derived from these observations

the length of the polar pendulum	= 39,217
while the length of the equatorial pendulum	= 39,017

hence the difference between the length of the seconds' pendulum at the pole and the equator amounts to 0·2 inches, which is the $\frac{1}{193}$th part of the whole length of the seconds' pendulum at the equator; but the length of the seconds' pendulum is a true representation of the intensity of gravity, the one being proportional to the other, since $g=\pi^2 L$, and therefore the decrease of the force of gravity for the quadrant is equal to about $\frac{1}{193}$th part of its value at the equator. Let us compare this result with the theoretical determination of the centrifugal force of rotating bodies.

A knowledge of the first elements of geometry and mechanics will enable the reader to understand the annexed proof of the theorem that 'the centrifugal force equal and opposite to the centripetal is directly proportional to the square of the velocity, and inversely proportional to the radius of the circle described,' or, expressed analytically,

$$f=\frac{v^2}{r}.$$

Let $OO'D$ (*fig.* 6) be a circle described by a body in uniform motion, and let OO' be an indefinitely small arc passed over in the unit of time, and therefore representing the velocity of the body. Let OO', which may be considered straight, be resolved into two components, the tangential OP and the central OQ, which form the sides of the rectangle $OPO'Q$, the diagonal of which is OO'. Joining O' with the extremity

Fig. 6.

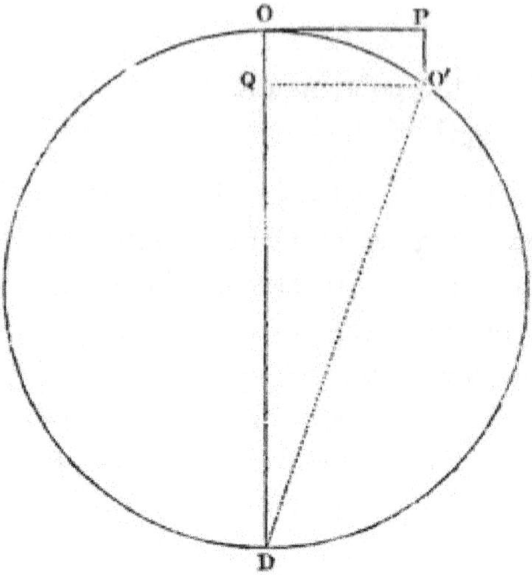

of the diameter D, we have the right-angled triangle OO′D, whence, by Euclid. Prop. VIII. Lib. 6:

$$OQ : OO' = OO' : OD,$$

$$\text{but } OQ = \tfrac{1}{2}f,\ OO' = v, \text{ and } OD = 2r$$

$$\therefore\ \tfrac{1}{2}f : v = v : 2r$$

$$\therefore\ f = \frac{v^2}{r}.\ \text{(Q. E. D.)}$$

Or, if the periodic time required to describe an entire revolution be expressed by t, then, since the circumference of the circle $= 2r\pi$, we have

$$v = \frac{2r\pi}{t}.$$

Substituting this value of v in the preceding formula, we have

$$f = 4\pi^2\frac{r}{t^2};$$

i.e. the centrifugal force is directly proportional to the radius of the circle described, and inversely proportional to the square of the periodic time. Applying this theorem to the rotatory motion of the earth, we shall find its centrifugal force at the equator by taking the equatorial radius $r = 20{,}923{,}596$ feet, and the number of seconds in a sideral year $t = 86{,}164''$, and, substituting these values in the last formula, we obtain, as the final result,

$$.f = 0{,}11126.$$

Let G denote the attraction of the earth at the equator (not influenced by rotation), and g the actual gravity at the same place, then

$$G = g + f;$$

but it has been found by observations made in the island of Rawak, that

$$g = 32{,}088 \text{ feet}$$

$$\therefore\ G = 32{,}088 + 0{,}11126 = 32{,}19926,$$

$$\text{whence } \frac{G}{f} = \frac{32{,}19926}{0{,}11126} = 289{,}4,$$

or, expressed in whole numbers, the force of the earth's attraction at the equator is 289 times the centrifugal force at the same place.

At first sight, there seems to be a discrepancy, since the decrease of gravity (from the poles to the equator) exceeds the centrifugal force (at the same place) by

$$\frac{1}{195} - \frac{1}{289} = \frac{1}{599}$$

very nearly; a small quantity in itself, but far too large as compared with the others in question not to be accounted for. A little consideration will convince us that this difference, instead of proving fatal to our hypothesis of the earth's rotation, tends to confirm it.

The rotation of the earth gives rise to the centrifugal force, the centrifugal force produces an ellipticity in the form of the earth itself, and this ellipticity modifies its power of attraction on bodies placed at its surface. Thus the same cause, viz., the rotation of the earth, exercises a direct and an indirect influence. The latter, viz., the oblate figure of the earth, has been deduced and calculated by Newton, Huyghens, and Clairaut, from the theory of gravitation and rotatory motion, and also by Laplace from the theory of the moon, and from precession and nutation; but it is not necessary for our purpose to enter into their abstruse calculations, or to examine the comparative merits of their analytical results, it is sufficient for us to state that the spheroidal form of the earth has been proved by the actual measurement of arcs of the meridian in different latitudes. We may cite as examples of the most important geodetical operations the French arc of 12° 22′ 12″, extending from Dunkirk to the Balearic Isles, and the English arc from Dunnose to Burleigh Moor, amounting to 3° 57′ 13″, and both the English and French measurements, when compared with those of Bouguer in South America, Lambton in India, and Svanberg in Lapland, give a general ellipticity of about $\frac{1}{300}$, which is probably as near the truth as local inequalities will admit. But it is evident that, the earth being elliptical, the attraction exerted by it on bodies placed near the equator must be less than that exerted on bodies near the poles, which are flattened, and therefore nearer to the centre. This diminution of attraction, which is owing to the oblateness of the earth, has been investigated by many eminent geometers, and the results show that independent of the centrifugal force, gravitation, considered in the direction of the pole to the equator, ought to decrease by almost exactly the $\frac{1}{599}$th part, which, together with $\frac{1}{289}$th part due to centrifugal force, makes up the $\frac{1}{195}$th part, the whole quantity actually found.*

Such a near coincidence of theory and observation renders our hypothesis the more incontestable, as there is no other to account for these differences of terrestrial gravity, even roughly estimated. Before concluding this chapter, we may briefly allude to another geographical phenomenon, which may be explained by the earth's rotation, namely, the trade winds. The difference of temperature of the air over the torrid and frigid zones could only generate southerly or northerly currents, but, in consequence of the rotation, every particle of air transferred from the pole to the equator must slacken its rotatory motion, and therefore acquire a relative direction towards the west, and thus assume the character of permanent north-easterly or south-easterly winds. But as they approach the equator from both sides simultaneously, their easterly tendency ought to diminish, and, upon meeting, be destroyed; so that the

* According to Clairaut's theorem, we have the ellipticity, as measured by the quotient of the difference of the axes by the smaller, equal to $\frac{5}{2}$ times the ratio of the centrifugal force at the equator to the force of gravity diminished by the ratio of the whole increase of gravity to the equatorial gravity. By this means we obtain the value of

$$\text{the ellipticity} = \frac{5}{2} \times \frac{1}{289} - \frac{1}{195} = 0{\cdot}00352,$$

while its value found from the geodetic measures is $= 0{\cdot}00335$

Difference . . $= 0{\cdot}00017$

equatorial region should be comparatively calm and free from easterly winds. All these consequences are agreeable to observed facts, and may therefore be taken as circumstantial evidence that the cause alleged, namely, the rotation of the earth, is the true one.

CHAPTER V.

THE DESCENT OF BODIES FROM GREAT HEIGHTS.

A DIRECT and most convincing demonstration of the earth's diurnal motion is furnished by those experiments which establish the fact that balls truly turned and carefully dropped from a great height do not fall exactly in the perpendicular, but deviate towards the east. Newton, in 1679, was the first who proposed to prove the rotation in this practical manner. The deviation of which Newton speaks is not, as Lalande (tom. 1, § 1082) asserts, to be accounted for by the curvature of the globe, but by the principles of mechanics. For if the earth rotates, the velocity of rotation must be greater at the summit of a tower than at its base, and a body dropped from the former will move faster in an easterly direction throughout the whole of its descent than the base does, and therefore reach the ground a little in advance of the foot of the perpendicular. Dr. Hooke, who was secretary of the Royal Society at the time Newton suggested this method, and who was renowned for his many mechanical contrivances, undertook to perform the necessary experiments. On January 14, 1680, he reported the results to the Society, stating that the height from which he dropped the balls was twenty-seven feet, and that they fell to the south-east. It was resolved that these experiments should be repeated before some of the members, but we find no further mention of them in the Transactions, and we are therefore inclined to conclude that they were not attended with success, owing to the smallness of the height, which could scarcely have caused a deviation of $\frac{1}{4}$ line. The idea, which originated with a philosopher whom all Europe revered, remained barren for a period of 112 years, when Guglielmini, a geometer of Bologna, revived it, and commenced making experiments in the tower degli Asinelli, which is 300 feet high. After many unsuccessful trials and various alterations in his apparatus, he gave up working in the day because he discovered that the traffic caused the tower to vibrate; and selected nights when the air was calm. The liberation of the balls was effected by burning the threads by which they were suspended, and this was never done until they had entirely ceased to oscillate, as observed by means of a microscope.

The balls fell upon a cake of wax spread upon the ground, and the holes made by them nearly coincided, the greatest distance between two centres scarcely amounting to half a line. The persevering philosopher finally dropped a plumb-line in August 1791, but was obliged to wait till February 1792 before it came to a state of perfect rest. He then found that the mean deviation of the sixteen balls he had dropped was 7·4 lines east and 5·27 lines south. These results were received by the scientific world with the greatest interest, when Lalande announced that Laplace's theory gave only a deviation of five lines to the east and none to the south. This striking difference of theory and observation was at once explained by a trifling circumstance, overlooked in the otherwise most accurate experiments of Guglielmini, but the explanation overthrew at the same time all their force of demonstration. The difference was occasioned, first, by a local disturbance of the air, owing to the numerous perforations in the walls of the tower, and secondly, by a slight curvature of its structure. The balls were dropped in the summer, but the vertical was only ascertained six months afterwards, in the winter, and there is no doubt that the difference of temperature was sufficient to shift the point of suspension. Although these experiments did not achieve the result for which they were undertaken, yet they were the precursors of others which furnished most satisfactorily the required proofs. Dr. Benzenberg commenced in 1802 a series of experiments in St. Michael's tower in Hamburgh, which is 402 feet * in height and most suitable for the purpose, for an uninterrupted fall of 340 feet may be obtained, exceeding that in the Asinelli by ninety-nine feet and that in St. Paul's by eighty-five feet. But as only 240 feet of the tower, from the base upwards, were completely closed, Benzenberg chose a height of 235 feet for his investigations, in order that even strong winds might not influence the result. The balls he used were 1·6 inches in diameter, and made of equal parts of lead and tin with the addition of a little zinc, an alloy which is easily turned and polished; they were tested by being placed upon quicksilver, and those which betrayed eccentricity in floating were rejected. His mode of suspension is represented in *fig.* 7. The suspending thread was cut, for burning might have caused a slight draught and an oscillation of the ball before being detached; and though he could not obtain permission to work at night, yet the vibrations of the tower were not perceptible, and

* Throughout this chapter, unless otherwise stated, the lengths are given in the French measure, viz. 1 foot = 12·78933 English inches.

Fig. 7.

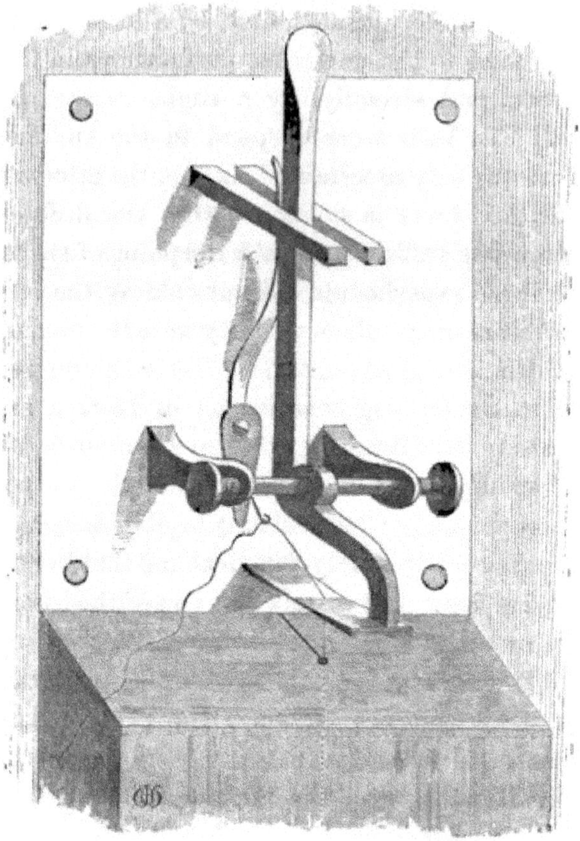

the weather proving very favourable, his experiments may be looked upon as accurate. The last thirty-one balls he dropped fell as follows:—

21	towards the east
8	,, ,, west
2	neutral
16	towards the south
11	,, ,, north
4	neutral

The special results of these thirty-one experiments will be seen in the following tables:—

	North	South	East	West
July 23, 1802	5	—	—	5
" "	—	5	8·5	—
" "	—	—	—	—
" "	1	—	—	—
" "	—	3	21·5	—
August 14 "	—	—	—	6
" "	—	—	16	—
" "	—	2·8	1·8	—
" "	6·8	—	5·2	—
August 16 ,, ...	—	1·7	—	10
" "	2·2	—	12	—
" "	—	0·8	—	1·5
" "	2	—	—	1
September 17	4	—	—	8
" "	8	—	7	—
" "	—	9	1·5	—
October 14 "	5·7	—	0·5	—
" "	—	17	12	—
October 15 "	—	10·5	17·3	—
" "	—	2	7	—
" "	—	8	0·8	—
" "	—	2	17·5	—
" "	0·2	—	14·7	—
" "	—	16·7	0·2	—
October 23 "	—	—	—	14·8
" "	—	5	8	—
" "	—	5·6	—	4·2
October 26 "	1	—	6	—
" "	—	5	13·5	—
" "	10·5	—	1·8	—
" "	—	0·5	1·7	—
Sum	46·4	92·6	174·5	50·5

South =	92·6	East =	174·5
North =	46·4	West =	50·5
	46·2		124·0

These numbers divided by 31 give respectively a mean deviation of four lines to the east and 1½ lines to the south. It will be seen that the first sixteen trials gave a mean deviation of 0·53 north and 2·6 east, while the last fifteen gave a deviation of 3·53 south and 5·46 east. This difference is explained by the fact that the apparatus was reversed, that is to say, the nippers which first opened towards the south afterwards opened towards the north, and at the same time the position of the suspending hook was changed from the west to the east. As this reversion seems necessary to avoid constant errors, we may accept the given deviations as fair results; but how do these agree with theory? We have already remarked, that the theory of Laplace does not give any southerly deviation, nor do the fundamental formulæ of Gauss allow any. Our object now, therefore, will be to investigate the question on physico-mathematical principles. In the preceding chapter we have seen that if the radius of the earth be denoted by r, and its periodic time of rotation by τ seconds, we have for centrifugal force at the equator

$$f = 4r\left(\frac{\pi}{\tau}\right)^2.$$

At any other point of the earth's surface the effect of the centrifugal force must be resolved into two components, the one acting in the prolongation of the radius, the other perpendicular to it.

Let the circle P O P′ represent any meridian of the earth, O any place on it, and let its latitude O E be denoted by L. Draw O M perpendicular to the axis of rotation P P′ and produce it till O N = O M′, then by the last formula the whole effect of rotation at O is expressed by

$$4\text{ON}\left(\frac{\pi}{\tau}\right)^2.$$

Joining O with the centre C producing C O, drawing a

FIG. 8.

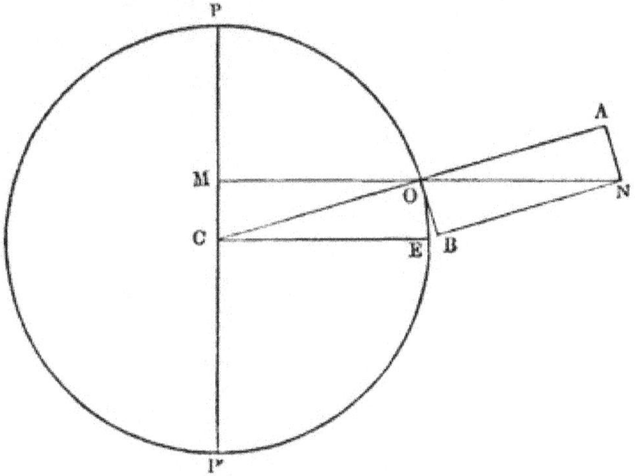

tangent at O and completing the parallelogram O A N B, we have

$$4\text{OA}\left(\frac{\pi}{\text{T}}\right)^2.$$

$$4\text{OB}\left(\frac{\pi}{\text{T}}\right)^2.$$

the respective expressions of the two components of the centrifugal force at O, but

$$\text{O A} = r\cos^2 \text{L} \text{ and } \text{O B} = r\sin \text{L} \cos \text{L};$$

hence by substitution, the component opposite, to the direction of gravity, and employed to diminish it, is expressed by

$$4r\cos^2 \text{L}\left(\frac{\pi}{\text{T}}\right)^2$$

and the other perpendicular to it in the plane of the meridian is expressed by

$$4r\cos \text{L} \sin \text{L}\left(\frac{\pi}{\text{T}}\right)^2 = 2r\ \sin 2\text{L}\left(\frac{\pi}{\text{T}}\right)^2.$$

If, therefore, from a point vertically above any place, the

latitude of which is L, and at which the actual gravity is g, a plumb-line be dropped, it will hang in a direction different from the vertical, and ρ, its angle of deviation from the perpendicular, will be determined by the equation

$$\rho = \frac{2r \sin 2 \text{ L } \left(\frac{\pi}{\tau}\right)^2}{g}.$$

Let the height from which the plumb-line is dropped be denoted by h, then δ, its southerly deviation, will be expressed in linear units by

$$\delta = \frac{2hr \sin 2 \text{ L } \left(\frac{\pi}{\tau}\right)^2}{g}.$$

Now, in order to find the southerly deviation, δ', of the point at which a ball dropped from the height h strikes the earth, let us assume that the falling body remains in the plane of a great circle perpendicular to the meridian (which must be the case if the resistance of the air is not taken into account), then this great circle will touch the small one of rotation of the place in question. If t denote the time of descent of the ball, then the angle of rotation for that time is

$$\varepsilon = 2\pi \frac{t}{\tau}.$$

Let P be the complement of the point o, and let the point where the ball strikes the earth be expressed by P′; then

$$\tan \text{P}' = \frac{\tan \text{P}}{\cos \varepsilon};$$

therefore

$$\tan (\text{P}' - \text{P}) = \left(\frac{\tan \text{P}}{\cos \varepsilon} - \tan \text{P}\right) : \left(1 + \frac{\tan^2 \text{P}}{\cos \varepsilon}\right)$$

$$= \frac{\tan \text{P}\,(1 - \cos \varepsilon)}{\cos \varepsilon + \tan^2 \text{P}} = \frac{2 \tan \text{P} \sin^2 \frac{1}{2}\varepsilon}{\cos \varepsilon + \tan^2 \text{P}}.$$

Now as ε in these experiments only amounts to about 36″, we may take $\cos \varepsilon = 1$; thus we have

$$\tan(\text{P}' - \text{P}) = \frac{2 \tan \text{P} \sin^2 \frac{1}{2} \varepsilon}{\sec^2 \text{P}} = \sin 2\,\text{P} \sin^2 \tfrac{1}{2} \varepsilon;$$

$$\text{but } \sin 2\,\text{P} = \sin 2\,\text{L},$$

$$\text{hence } \tan(\text{P}' - \text{P}) = \sin 2\,\text{L} \sin^2 \tfrac{1}{2} \varepsilon.$$

No appreciable error can arise by taking the arc $(\text{P}' - \text{P})$ for $\tan(\text{P}' - \text{P})$, whence we obtain the southerly deviation expressed in linear measure,

$$\delta' = r \sin 2\,\text{L} \sin^2 \tfrac{1}{2} \varepsilon.$$

In order to eliminate ε, we take the equation $\varepsilon = 2\pi \frac{t}{\text{T}}$, from which we derive

$$\sin^2 \tfrac{1}{2} \varepsilon = \pi^2 \frac{t^2}{\text{T}^2};$$

therefore, by substitution,

$$\delta' = rt^2 \sin 2\,\text{L} \left(\frac{\pi}{\text{T}}\right)^2,$$

and if the ball fall in vacuo

$$t^2 = \frac{h}{\frac{1}{2} g}$$

whence we have $\delta' = \dfrac{2r\,h \sin 2\,\text{L} \left(\frac{\pi}{\text{T}}\right)^2}{g}$

which is identical with the expression for δ, as determined above.

From this it will be obvious that if there is no resistance of the air, there is no southerly deviation, but as it cannot be neglected,

$$t^2 > \frac{h}{\frac{1}{2} g};$$

$$\text{therefore } \delta' > \delta,$$

whence it would appear that a deviation exists. We must emember, however, that the expression for δ' was found by assuming that the ball never quits a certain plane of a great circle: now this cannot be true in a resisting medium, for the conical motion of the atmosphere will rge the ball towards the north, just sufficiently to compensate for the apparent difference $\delta'-\delta$. This may be proved in the following manner. If in the formula

$$\delta' = rt^2 \sin 2\,\text{L} \left(\frac{\pi}{\text{T}}\right)^2$$

we substitute $\tan \rho = \dfrac{2r \sin 2\,\text{L} \left(\frac{\pi}{\text{T}}\right)^2}{g}$

we have $\delta' = \frac{1}{2}\, gt^2 \tan \rho$.

Now as $\frac{1}{2}\, gt^2$ expresses the height through which the ball would fall in vacuo during t seconds, it follows that if there be no resisting medium the ball must pass in the direction of the plumb-line; but owing to the resistance of the air, the space of descent is not $\frac{1}{2}\, gt^2$ but h, and hence,

$$\delta' = h \tan \rho = \frac{2hr \sin 2\,\text{L} \left(\frac{\pi}{\text{T}}\right)^2}{g};$$

we have therefore again the result

$$\delta' = \delta,$$

i. e., no southerly deviation can exist, even if the experiment be made in air.

With regard to the easterly deviation, we know that a ball dropped from a height h has a velocity caused by the rotation of the earth

$$\text{V} = \frac{2\pi\,(r+h)\cos \text{L}}{\text{T}}$$

in the direction of the tangent to the circle of latitude of the place of observation, with such a velocity that the ball would describe a space

$$vt = \frac{2\pi t (r + h) \cos L}{\tau},$$

while the point of the earth vertically below the place of observation passes through an arc the length of which is expressed by

$$\frac{2\pi tr \cos L}{\tau},$$

and as arc and tangent have in this case no appreciable difference, we obtain for the easterly deviation

$$\frac{2\pi th \cos L}{\tau}.$$

This is true if the directions of gravity towards the falling body remain parallel, and, though their angles of inclination are small, they must not be neglected.

Fig. 9.

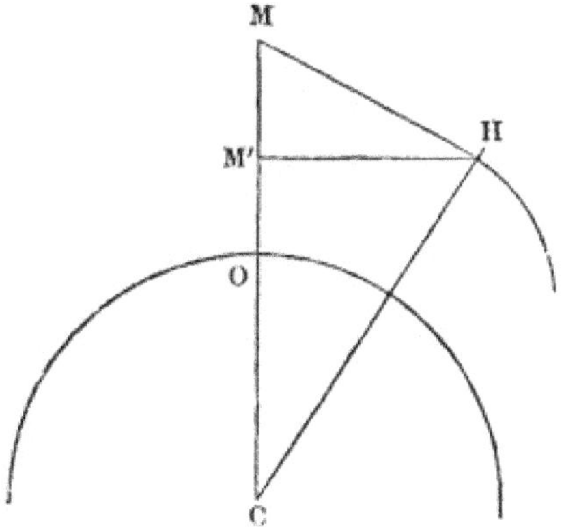

Suppose the body during its descent to have arrived

at H, and let $M'H = x$, let w denote its velocity in the direction $M'H$, and v that in the direction HC, then

$$dv = gdt,$$

$$\text{therefore } dw = \frac{-gx \cdot dt}{r}$$

$$\text{taking } HC = MC = r.$$

$$\text{But } w = \frac{dx}{dt} \text{ or } dt = \frac{dx}{w};$$

whence by substitution

$$dw = -\frac{gx\,dx}{wr}$$

$$\text{or } wdw = -\frac{gx\,dx}{r},$$

which by integration gives

$$w^2 = -\frac{gx^2}{r} + \text{constant};$$

and since for $x = o$ the original tangential velocity is $V = \frac{2\pi}{\tau}(r + h)\cos L$, we have by correction

$$w^2 = V^2 - \frac{gx^2}{r},$$

$$\text{or } w = V\sqrt{\left(1 - \frac{gx^2}{rV^2}\right)},$$

Taking again the equation

$$dt = \frac{dx}{w} \text{ we have by substitution}$$

$$Vdt = \frac{dx}{\sqrt{1 - \frac{gx^2}{rV^2}}} = dx\left(1 + \tfrac{1}{2}\frac{gx^2}{rV^2}\ \ldots\ldots\right).$$

Neglecting the other terms of the series, we obtain by integration,

$$Vt = x + \tfrac{1}{6}\frac{gx^3}{rV^2},$$

no correction being necessary for $x = o$ when $t = o$. It is also evident that the second term of the right side of the equation is very small compared with x, and we may therefore without sensible error substitute in it $\mathrm{v}^3 t^3$ for x^3, and write

$$\mathrm{v}t = x + \tfrac{1}{6}\frac{g\mathrm{v}t^3}{r}$$

$$\text{or } x = \mathrm{v}t - \tfrac{1}{6}\frac{g\mathrm{v}t^3}{r}$$

Now if $\triangle$ denote the excess of descent in vacuo above that in air, we have

$$\tfrac{1}{2} gt^2 = h + \triangle,$$

$$\text{also } \mathrm{v} = \frac{2\,\pi}{\tau}\,(r + h)\,\cos \mathrm{L}$$

$$\text{and } \frac{r + h}{r} = 1,$$

which is true for the 0·000001 part of a line; therefore

$$x = \frac{2\pi}{\tau}\,t\,(r + h)\,\cos \mathrm{L} - \frac{2\pi}{3\tau}\,t\,(h + \triangle)\,\cos \mathrm{L};$$

and because the plane 0 advances towards the east in t seconds by

$$\frac{2\pi}{\tau}\,tr\,\cos \mathrm{L},$$

we get for the easterly deviation of the fallen ball the expression

$$\frac{2\pi}{\tau}\,th\,\cos \mathrm{L} - \frac{2\pi}{3\tau}\,t\,(h + \triangle)\,\cos \mathrm{L}$$

$$= \frac{4\pi\,t\,(h - \tfrac{1}{2}\triangle)\,\cos \mathrm{L}}{3\tau}.$$

Applying this formula to Benzenberg's experiments, we must take $\mathrm{L} = 53° 33'$, the latitude of Hamburg, and

$h = 235$ feet, from which the other quantities t and Δ may be calculated, $t = 4'' \, 1'''$, $\Delta = 8{\cdot}787$, while τ and π are known as constants, $\tau = 86164''{\cdot}1$, and $\pi = 3{\cdot}14$ . .

The calculation is rendered very simple by the use of logarithms, as will be seen by the following:—

Calculation of Experiments made in St. Michael's Tower, Hamburg, anno 1803, from a height of 235 feet, and under 53° 33′ latitude, by Dr. Benzenberg.

$$\text{Formula.} \quad x = \frac{4\pi \, t \, (h - \frac{1}{2}\,\Delta) \cos \text{L}}{3\tau}.$$

$$h = 235$$
$$t = 4'' \, 1''' = 4''{\cdot}017$$
$$\tfrac{1}{2}\, g = 15'{\cdot}108$$
$$h + \Delta = 243'{\cdot}787 = \tfrac{1}{2}\, gt^2$$
$$\Delta = 8'{\cdot}787$$
$$h - \tfrac{1}{2}\,\Delta = 230'{\cdot}607$$
$$\text{L} = 53^\circ \, 33' \quad \tau = 86164''{\cdot}1$$

log. 144 = 1·158362 . . (The factor 144 is inserted to reduce the feet into lines.)

log. 4π = 1·099212

log. t = 0·603902

log. $(h - \frac{1}{2}\,\Delta)$ = 2·362872

log. cos L = 9·773875 − 10

5·998223

− log. 3τ = 5·412447

0·585776

No. of which = 3·852 lines, result of theory.

3·997 „ „ observation.

Difference 0·145

This result shows how, as regards the easterly deviation,

theory and observation agree, but there still remains a southerly deviation of $1\frac{1}{2}$ lines which is not accounted for by theory. Benzenberg, not quite satisfied with the result of his experiments, determined to repeat them, and chose for this purpose an abandoned coal-pit at Schlebusch, in the county of Mark, Westphalia, situated in a latitude 51° 25′, and about 264 feet deep, of which 262 feet were available for the free descent of the balls. With these data, and taking $\frac{g}{2} = 15{\cdot}105$ and $t = 4''{\cdot}164$, we get by the formula

$$x = \frac{4\pi\, t\,(h - \frac{1}{2}\,\Delta)\cos \text{L}}{3\text{T}}$$

an easterly deviation of 4·64 lines. The following is the calculation:—

$$\begin{aligned} \log. 4\pi &= 1{\cdot}099212 \\ \log. t\,[4''{\cdot}164] &= 0{\cdot}619510 \\ \log. h - \tfrac{1}{2}\,\Delta\,[262 - 5{\cdot}36] &= 2{\cdot}406926 \\ \log. 144 &= 2{\cdot}158363 \\ \log. \cos \text{L}\,[51^\circ\, 25'] &= \underline{9{\cdot}794942 - 10} \\ & \quad 6{\cdot}078953 \\ \log. 3\text{T} &= \underline{5{\cdot}412441} \\ & \quad \underline{\underline{0{\cdot}666512}} \end{aligned}$$

the number of which = 4·64 lines

Having now examined the theory, we will turn our attention to Benzenberg's new experiments, twenty-nine in number. They were commenced in the autumn of 1804, all precautionary arrangements having been made, such as covering the entrance of the pit and blocking up its lower outlets.

	North	South	East	West
October 7, 1804	11·5	—	—	3
" "	14·5	—	12	—
" "	5	—	3	—
October 8 "	4	—	13	—
" "	1·5	—	20	—
" "	—	2	—	2
" "	—	9	11·5	—
" "	—	0·5	—	4
" "	—	15	2	—
" "	8	—	2	—
" "	—	5	12	—
" "	13·5	—	7	—
" "	3·5	—	13·5	—
" "	—	0·5	11	—
October 9 "	13	—	9	—
" "	—	4	—	8
" "	8	—	8	—
" "	10	—	10	—
" "	—	13	7	—
" "	—	11	7·5	—
" "	19	—	6	—
" "	—	8·5	—	2
" "	—	0·5	11	—
" "	—	15	—	4
" "	—	6	—	9
" "	—	7	—	10
" "	8	—	8·5	—
" "	—	6	10	—
" "	4·5	—	5·5	—
Sum	124	103	189·5	42

North =	124	East =	189·5
South =	103	West =	42
Difference	21	Difference	147·5

These numbers equally distributed among the twenty-nine trials, give as a mean result for a single ball,

Northerly deviation = 0·7 lines
Easterly deviation = 5·1 "

We see from this that no deviation took place towards the south and the small amount, viz. $\frac{7}{10}$ of a line northerly deviation, is obviously due to the errors which can never be avoided in practice. On the other hand, the deviation towards the east given by observation approaches so closely to the theoretical result 4,6 calculated upon the hypothesis of the earth's rotation, that the latter is confirmed, and the more so as no other reason can be assigned for the aggregate deviation of all the balls towards the east exceeding half the sum of the eastern and western deviations. An additional and very strong confirmation has since been furnished by the favourable result of a long series of similar experiments made by Professor Reich in the year 1831 in the mines of Freiberg. The depth he could command for the free descent of the balls was 488 feet, nearly double that of Schlebusch, the number of trials he made amounted to 106, and the mean result gave only a slight southerly deviation, viz. less than half a line, and an easterly deviation of 10,32, which differs in defect by 1,85 from the results of theory.

Calculation of Experiments made in the Mines of Freiberg, from a height of 488,06 feet, and under 50° 53′ 23″ latitude, by Professor Reich.

$$\text{Formula.}\quad x = \frac{4\pi\, t\,(h - \frac{1}{2}\,\Delta)\cos \mathrm{L}}{3\tau}.$$

$$h = 488{\cdot}06$$

$$t = 6''{\cdot}0098$$

$$\tfrac{1}{2}\,g = 15{\cdot}097$$

$$h + \Delta = 545{\cdot}317$$

$$\Delta = 57{\cdot}257$$

$$h - \tfrac{1}{2}\,\Delta = 459{\cdot}432$$

$$\mathrm{L} = 50^\circ\ 53'\ 23''. \quad \tau = 86164''{\cdot}1$$

$$\begin{aligned}
\log. 144 &= 2{\cdot}158362 \\
\log. 4\pi &= 1{\cdot}099212 \\
\log. t &= 0{\cdot}778216 \\
\log. (h - \tfrac{1}{2}\Delta) &= 2{\cdot}662217 \\
\log. \cos \angle &= 9{\cdot}799902 - 10 \\
&\overline{6{\cdot}497909} \\
-\log. 3\tau &= 5{\cdot}412447 \\
&\overline{1{\cdot}085462}
\end{aligned}$$

the number of which is = 12·17 lines.

The following table shows the results of 106 experiments, expressed in millimètres: —

FIRST SERIES OF 18 EXPERIMENTS.

	No.	South	North	East	West
August 23 . . .	1	—	61, 8	38, 7	—
„ 25 . . .	7	—	102, 8	179, 0	—
„ — . . .	13	84,15	—	—	77,05
„ — . . .	19	85, 0	—	—	12,85
„ 27 . . .	3	—	129, 3	11,55	—
„ — . . .	9	3,85	—	16,65	—
„ 29 . . .	3	41,55	—	—	80,15
„ — . . .	9	—	129,65	60,95	—
Sept. 1 . . .	3	—	2, 6	56, 6	—
„ 1 . . .	9	—	16,55	146, 6	—
„ 1 . . .	15	—	8, 3	67,85	—
„ 6 . . .	3	—	71, 7	96,05	—
„ — . . .	9	13, 2	—	4, 6	—
„ — . . .	15	—	1, 7	—	57,95
„ — . . .	21	—	34, 9	—	7,15
„ 7 . . .	6	—	147,55	47,19	—
„ — . . .	12	193, 8	—	94, 4	—
„ — . . .	18	—	130, 9	—	19, 9
		421,55	837,75	820,14	255,05

SECOND SERIES OF 18 EXPERIMENTS.

	No.	South	North	East	West
August 24	2	6, 3	—	49, 3	—
,, 25	8	—	56,95	37, 7	—
,, —	14	19,65	—	60, 3	—
,, —	20	—	51, 9	52, 7	—
,, 27	4	—	97,55	—	65,35
,, —	10	65,15	—	—	105,85
,, 29	4	—	6, 2	—	58, 6
,, —	10	65, 5	—	—	61,85
Sept. 1	4	62,85	—	72,65	—
,, 1	10	27, 7	—	46,10	—
,, 1	16	—	22, 9	—	38, 7
,, 6	4	47, 4	—	—	6, 3
,, —	10	62, 1	—	62, 5	—
,, —	16	26, 5	—	58,15	—
,, 7	1	—	97, 2	41,09	—
,, —	7	3, 0	—	127,19	—
,, —	13	117,95	—	48,65	—
,, —	19	104, 0	—	—	163, 3
		608,10	332,70	656,33	499,95

THIRD SERIES OF 18 EXPERIMENTS.

	No.	South	North	East	West
August 24	3	—	5,05	—	37,15
,, 25	9	—	123, 1	—	18,75
,, —	15	79,85	—	—	9,52
,, —	21	123, 0	—	—	4, 5
,, 27	5	15, 6	—	—	14, 4
,, —	11	11,85	—	—	41,15
,, 29	5	97,55	—	66, 8	—
,, —	11	26,45	—	41, 6	—
Sept. 1	5	—	31,05	—	45,45
,, 1	11	13, 8	—	92, 9	—
,, 1	17	19, 6	—	85, 8	—
,, 6	5	14, 9	—	45, 4	—
,, —	11	—	11,85	69,15	—
,, —	17	—	4, 8	61, 4	—
,, 7	2	66, 9	—	—	141,81
,, —	8	—	8, 9	68,24	—
,, —	14	—	114,65	32, 8	—
,, —	20	—	31, 9	3, 8	—
		469,50	332,30	567,49	312,73

FOURTH SERIES OF 18 EXPERIMENTS.

	No.	South	North	East	West
August 24 . . .	4	62, 8	—	—	60,75
„ 25 . . .	10	—	12,45	—	36,95
„ — . . .	16	18, 5	—	63, 7	—
„ — . . .	22	—	109,95	19,55	—
„ 27 . . .	6	184,35	—	75, 8	—
„ — . . .	12	11, 7	—	90, 0	—
„ 29 . . .	6	—	79, 8	29, 3	—
„ — . . .	12	110,55	—	79,05	—
Sept. 1 . . .	6	—	0,75	3,75	—
„ 1 . . .	12	33,15	—	60,15	—
„ 1 . . .	18	18, 9	—	37,35	—
„ 6 . . .	6	—	99, 4	57,85	—
„ — . . .	12	35,75	—	42, 5	—
„ — . . .	18	—	2, 9	6, 0	—
„ 7 . . .	3	—	39, 6	—	45,31
„ — . . .	9	25, 5	—	99,64	—
„ — . . .	15	34, 2	—	—	30,95
„ — . . .	21	—	151, 2	27, 0	—
		535,30	496,05	691,64	173,96

FIFTH SERIES OF 17 EXPERIMENTS.

	No.	South	North	East	West
August 24 . . .	5	6,65	—	48, 1	—
„ 25 . . .	11	21, 3	—	69,45	—
„ — . . .	17	—	26, 5	—	6, 6
„ 27 . . .	1	—	11,65	—	9,75
„ — . . .	7	141,1	—	88, 1	—
„ 29 . . .	1	—	86, 0	17,55	—
„ — . . .	7	33, 0	—	24,95	—
Sept. 1 . . .	1	44,55	—	—	43,15
„ 1 . . .	7	70,05	—	22, 8	—
„ 1 . . .	13	1,15	—	33, 9	—
„ 6 . . .	1	—	86,85	—	23, 9
„ — . . .	7	—	20,45	30,15	—
„ — . . .	13	—	38, 3	—	34, 4
„ — . . .	19	—	27,85	12, 8	—
„ 7 . . .	4	—	27, 2	10,79	—
„ — . . .	10	—	130, 8	57,94	—
„ — . . .	16	8, 1	—	15, 0	—
		325,95	455,60	431,53	117,80

Sixth Series of 17 Experiments.

	No.	South	North	East	West
August 24 . . .	6	83, 6	—	—	40,75
,, 25 . . .	12	11, 6	—	—	5,55
,, — . . .	18	95, 2	—	74,95	—
• ,, 27 . . .	2	—	7, 3	6, 5	—
,, — . . .	8	88, 8	—	119,15	—
,, 29 . . .	2	—	128, 7	49, 0	—
,, — . . .	8	39,45	—	30, 9	—
Sept. 1 . . .	2	—	30,65	90, 9	—
,, 1 . . .	8	72,55	—	49,75	—
,, 1 . . .	14	—	26,65	94,35	—
,, 6 . . .	2	—	34, 9	46,55	—
,, — . . .	8	18, 9	—	95, 0	—
,, — . . .	14	—	20,25	32,95	—
,, — . . .	20	25, 2	—	55, 0	—
,, 7 . . .	5	—	104,75	—	134,66
,, — . . .	11	117, 9	—	29, 6	—
,, — . . .	17	—	21,95	57,85	—
		553,20	375,15	832,45	180,96

Final Results.

1. Series	421,55	837,75	820,14	255,05
2. ,, ...	608,10	332,70	656,33	499,95
3. ,, ...	469,50	332,30	567,49	312,73
4. ,, ...	535,30	469,05	691,64	173,96
5. ,, ...	325,90	455,60	431,53	117,80
6. ,, ...	553,20	375,15	832,45	180,96

From which we derive—

1st. *For the northerly and southerly deviation.*

South — North

Millimètres

$$\frac{421{,}55 - 837{,}75}{18} = -\ 23{,}122.$$

$$\frac{608{,}10 - 332{,}70}{18} = +\ 15{,}300.$$

South — North

Millimètres

$$\frac{496,50 - 332,30}{18} = + \ 7,622.$$

$$\frac{535,30 - 469,05}{18} = + \ 3,686.$$

$$\frac{325,90 - 455,60}{17} = - \ 7,629.$$

$$\frac{553,20 - 375,15}{17} = + 10,473.$$

2nd. *For the easterly and westerly deviation.*

East — West

Millimètres

$$\frac{820,14 - 255,05}{18} = + 31,393.$$

$$\frac{656,33 - 499,95}{18} = + \ 8,687.$$

$$\frac{567,49 - 312,73}{18} = + 14,153.$$

$$\frac{691,14 - 173,96}{18} = + 28,760.$$

$$\frac{431,53 - 117,80}{17} = + 18,454.$$

$$\frac{832,45 - 180,96}{17} = + 38,323.$$

Mean Result.

North — South	East — West
37,081 − 30,751 = 6,330	+ 139,770

These results divided by six, the number of series, give the following mean deviations: —

Towards the south	Towards the east
$\frac{6,330}{6} = 1,055$ millim.	$\frac{139,770}{6} = 23,295$ millim.;

or, since one millim. = 0,4433 Paris lines, we have by reduction,

Southerly deviation = 0,46 Paris lines
Easterly deviation = 10,32 " "

We cannot conclude this chapter without expressing a wish that some scientific man, emulated by the truly practical spirit so often met with in Englishmen, may repeat these delicate and arduous experiments. This country offers singular local advantages for the prosecution of such researches. We have deep mines and high buildings, and we only require an enterprising and persevering philosopher to corroborate by his testimony the results of those inquiries which, though emanating originally here, have as yet only found successful votaries abroad.

CHAPTER VI.

ON THE CHANGE OF THE PLANE OF OSCILLATION OF THE PENDULUM.

THE pendulum, which has become so useful in science and in art (ever since the day when Galileo reflected on the swinging of a lamp) — the pendulum, which is the most perfect measure of time, which serves to determine the figure of the earth, which is so intimately connected with the progress of modern astronomy; this simple and beautiful instrument is again chosen in this investigation as the best auxiliary in physical researches. It is a well-known fact, that if any weight, attached to the end of a string, be suspended from a fixed point it will remain in equilibrum, when in a vertical position, but if drawn into any other and then liberated, it will descend by the action of gravity with accelerated motion, and by reason of its inertia continue to move, and, therefore, ascend with retarded motion; moreover, the swinging down and up would last for ever if the point of suspension had no friction, and the air offered no resistance. It is further self-evident, that if the pendulum be not acted upon by any other force than gravity — if it commenced with no initial velocity, and no extraneous impulse interfere, it must preserve its original plane of vibration. Should, on the other hand, the pendulum be exposed to a jerk, however slight, it will describe the surface of a cone with elliptical base. This elliptical motion was once made use of by Dr.

Hooke to establish a representation of the laws of nature in the solar system. It was one evening, in the year 1774, when he read before the Royal Society his paper on the laws of attraction, that, for sake of illustration, he suspended a ball from the ceiling, and, while the pendulum was vibrating, he communicated to it an impulse, laterally with respect to the plane of vibration; immediately the ball was found to describe an ellipse about a centre which was in the vertical line passing through the point of suspension.

He succeeded in some degree in explaining by this method the principle upon which the elliptical motion of the earth round the sun depends, but his experiment is imperfect, inasmuch as he assumes the sun to be in the centre instead of the focus of the terrestrial orbit; besides, it affords no demonstration of the earth's motion, for, in order to show its rotation by means of the pendulum, all elliptical swing must be carefully avoided. The ocular evidence of the earth's motion is founded upon the principle of mechanics, that the simple pendulum cannot, by its own agency, change its original plane of oscillation; if, therefore, any change take place it must be attributed to some external cause.

The Academicians del Cimento, at Florence, who made many experiments on the pendulum, established the fact that the plane of vibration varies. We may here quote two passages which occur in the 'Saggi di Naturali Experienze,' edizione del 1841, p. 20, and in the 'Notizie degli Aggrandimenti delle Scienze fisiche in Toscana,' vol. ii. p. 669, which show the antiquity of the discovery.

The first is as follows: —

'The simple pendulum with a single thread suspension

being at liberty to move round deviates farther and far from its original position until it comes to rest: it does not move in a vertical arc but follows an oval spiral.'

The other paragraph in the notes published by Targioni runs thus: —

'November 28, 1661.

'If the pointed bob of a simple pendulum with a single thread suspension be made to touch during its swing a plane strewed with sand, it will be observed that its vibrations cease to be in a vertical arc and follow a spiral curve, thus tracing in the sand an oval spiral the branches which all intersect in the centre.'

Notwithstanding that these quotations prove that the academicians of years gone by were aware of the deviation of the pendulum from its vertical plane, yet we have no evidence to show that they applied this discovery to the demonstration of the earth's rotation. It is to Mr. Foucault that the merit is due of having first shown the relation and dependence of these two motions. Mr. Foucault communicated a description of his experiments, on February 3, 1851, to the Academy of Sciences at Paris; his apparatus consists of a steel wire fastened at the upper extremity in a metal plate by means of a screw (see *fig.* 10) and carrying at the other end a heavy ball of brass, to which is attached a pointed index. In order to avoid any initial velocity, the pendulum is drawn up out of the vertical and held in that position by means of a thread passed round the bob and made fast to a fixed object; this being done, and the ball in a perfect state of rest, the thread is burnt through, the pendulum commences its swing in the vertical plane of its point of suspension;

it will soon, however, be observed to leave this and deviate more and more towards the west. We shall proceed hereafter to determine theoretically the amount of this deviation and compare it with the results of observation, but it will now suffice to remark that the motion of the pendulum

FIG. 10.

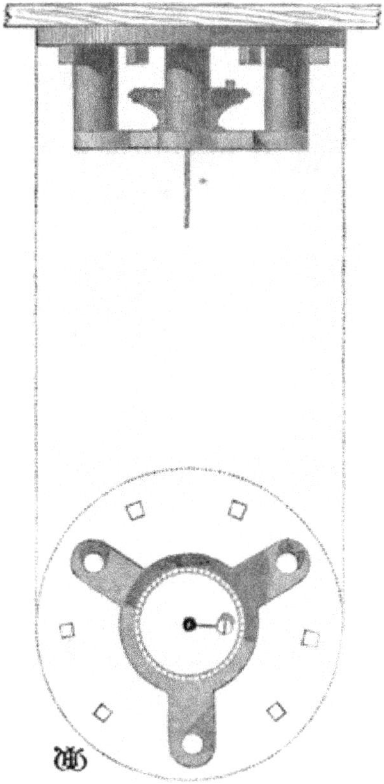

FOUCAULT'S SUSPENSION

in the azimuth is only apparent, that its plane of oscillation remains fixed, and that it is the earth which travels under it. Objection may be taken to this proof, inasmuch as it may be suggested that the point of suspension being fixed to the earth, participates in its motion, but the torsion of the wire thus occasioned is so small as to exercise no sensible influence on the whole. It is necessary, in order to have a clear conception of the fixity of the plane of suspension, to be thoroughly acquainted with the following two principles of mechanics: —

1st. If the pendulum commencing its swing in a vertical plane rotate at the same time round its fixed point of suspension then the oscillations will continue in the same plane.

2nd. If the centre of suspension rotate also, then the oscillations will take place in planes parallel to the first plane of motion.

In order to illustrate this tendency of the pendulum to move in the same plane or in one parallel to its original motion, I have used the following apparatus (see *fig.* 11). R R, represents a brass ring fitted in a wooden frame W W, and capable of being turned round by means of the handle H, the whole supported by a tripod stand; D is a disc of metal fastened to the wooden frame by a strip of brass, concentric with the ring but of smaller radius, so that ample space is left for the rotatory motion of the pendulum P P, bent at the upper part, which terminates in a small cone of hardened steel fitting into a corresponding cup drilled in the centre of the disc; Q Q represents a similar pendulum of smaller dimensions capable of free movement at the extremity E of a rod E E, bent at right angles and fastened to the brass ring.

If now the heavy bob of the pendulum, P P, be raised and carefully dropped, it will be observed to swing in a vertical plane; let its direction be marked by comparison with the position of a fixed object, and let a lateral

Fig. 11.

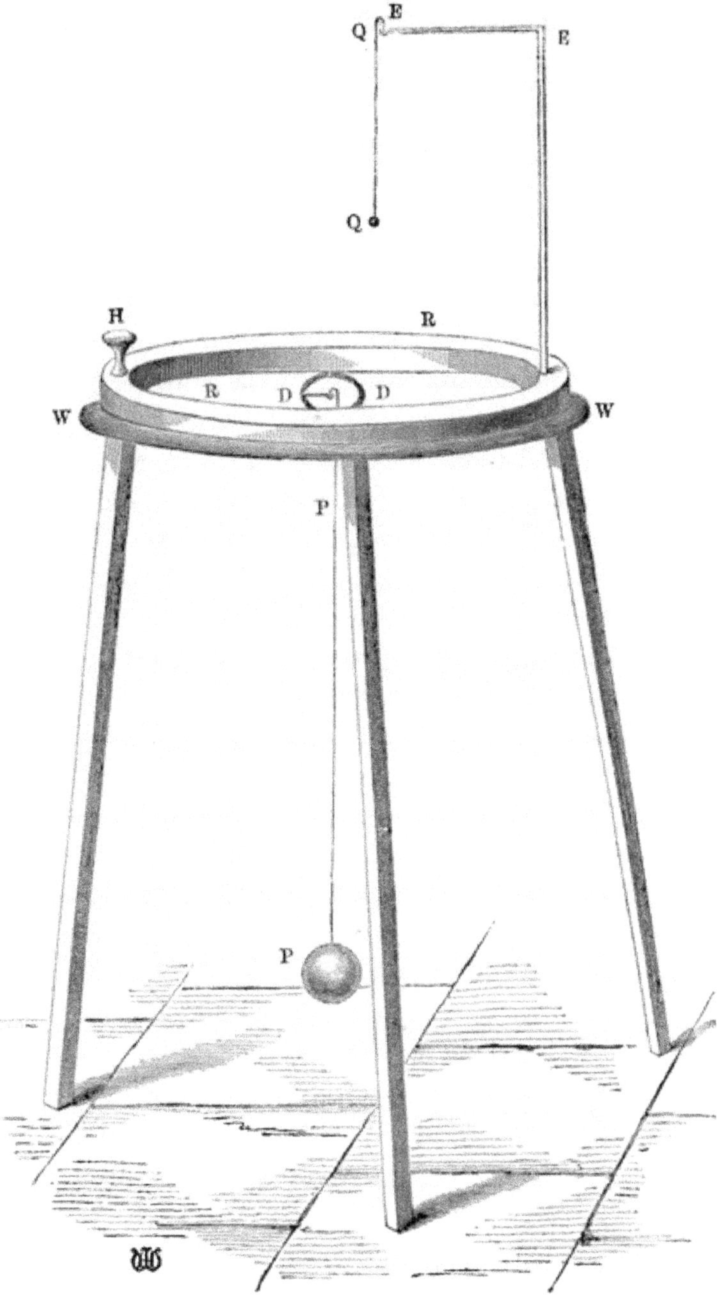

APPARATUS FOR SHOWING PARALLELISM OF PLANE OF OSCILLATION.

movement be repeatedly communicated to it so that it may rotate through arcs of different degrees about the disc D; then the following curious phenomenon will manifest itself, viz. the plane of oscillation remains unchanged. If we take into consideration that the motion of the earth can exercise no appreciable influence on the instrument in the short duration of the pendulum's motion, we may take this experiment as furnishing a convincing proof of our first proposition. In order to investigate the truth of the other proposition, we set the smaller pendulum, Q Q, swinging in a plane, for instance, in one parallel to the rod, S R, and turn the brass ring by means of the handle H steadily round, it will then be apparent that while the rod rotates the pendulum will continually shift about its point of suspension so as to swing parallel to its original plane of oscillation.

Having now demonstrated in a practical manner these two mechanical principles, we may proceed to investigate the effects upon the free pendulum which should result from the earth's rotation. Let us first imagine that the place for experiment is exactly over the North Pole. Let the pendulum be swung so as to coincide with the plane of the meridian of London; then, as the earth rotates, the point of suspension will remain fixed; but the motion of the mass originally impressed parallel to a given plane will continue in that plane, and consequently the latter will coincide in the course of twelve hours with every meridian in succession, and the apparent rotation of the pendulum from east to west will be entirely completed in twenty-four hours. In any other latitude between 90° and 0, a continuous and regular apparent change of motion must occur; for, though the point of suspension rotates, the planes of oscillation remain parallel to each other (at least, their intersections with the horizon); but a meridian line of the globe does not preserve its parallelism during diurnal motion, except at the equator; therefore, at all points of the earth's surface, except at the equator, we must observe this phenomenon, that the plane of motion varies uniformly, apparently remaining to the westward, while the earth rotates eastward quicker, however, in high than in low latitudes.

To find the angular velocity of the pendulum's apparent motion mathematically for any place, it is necessary to resolve the rotation of the earth round its axis into two directions, the horizontal and the vertical of the place of observation; one of which, namely, that round the vertical, is alone effective.

The method of resolution and composition of rotatory motion is analogous to that of rectilinear, founded on the theorem known in mechanics by the name of the parallelogram of forces, as may be proved in the following way:—

Suppose a body rotate about an axis C A with an angular

FIG. 12.

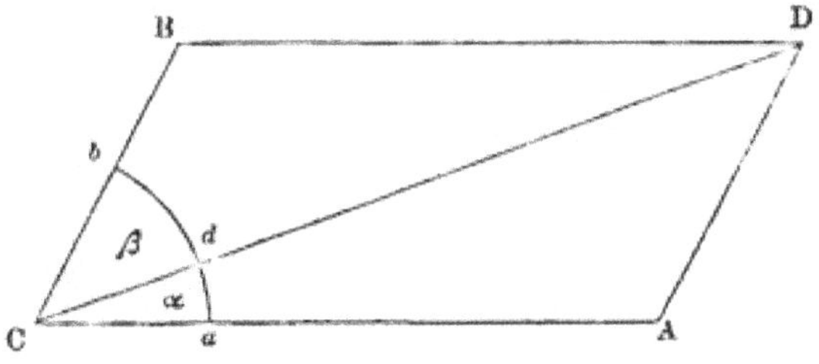

velocity p proportional to the length C A, and at the same time about another axis C B with an angular velocity q

proportional to the line CB, then the resultant motion will be a rotation about the diagonal CD of the parallelogram CBDA, having an angular velocity *n* proportional to the line CD, so that

$$p : q : n = CA : CB : DC = \sin \beta : \sin \alpha : \sin (\alpha+\beta)$$

With the centre C and a radius taken as unity, let a circle be described in the plane of the axes CB and CA intersecting the lines CA, CD, CB respectively in *a d b*, then the point *d* will, in consequence of its rotation about the axis CA, describe during the instant of time Γ a small arc perpendicular to the plane through the axis with a radius equal to sin α, so that its elevation above the plane becomes, after the lapse of Γ,

$$\Gamma p \sin \alpha$$

but, in consequence of the simultaneous rotation of the point *d* about the axis CD, it must undergo at the same time a depression expressed by

$$\Gamma q \sin \beta$$

but because

$$p : q = CA : CB = \sin \beta : \sin \alpha :$$

therefore

$$\Gamma q \sin \beta = \Gamma p \sin \alpha \ldots (1)$$

The same is true for any other point of the line CD, hence the diagonal CD is the resultant axis of rotation. Considering the point *b* we have for its elevation, in consequence of its original rotation about CA,

$$\Gamma p \sin (\alpha+\beta),$$

but in consequence of the resultant rotation we have for the same expression

$$\Gamma n \sin \beta$$

therefore:

$$p \sin(\alpha+\beta) = n \sin \beta$$

which gives by substitution in equation (1)

$$p : q : n = \sin \beta : \sin \alpha : \sin(\alpha+\beta)$$

This theorem may now be applied to determine the angular velocity with which the plane of oscillation of a freely-suspended pendulum moves in the azimuth from east to west. Let L be any place on the earth, and let C A represent the amount of rotation of the earth about its axis P P′,

FIG. 13.

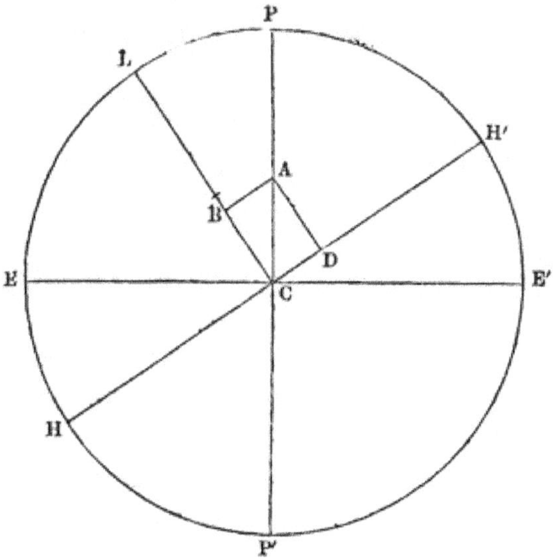

for a short interval of time, then drawing through L the vertical C L and the line H H′, perpendicular to it through the centre of the globe, we have by completing the rectangle A B C D, replaced the whole rotation C A by its two component parts C B and C D. But with regard to the rotation C D about the axis H H, the pendulum at L will undergo the same conditions as if placed at the equator, whence it follows that the direction of the plane of oscillation and the velocity with which it moves are not influenced by the component rotation C D about the axis H H. It must therefore be admitted that the velocity will be the same as if a rotation about C B alone took place. Hence we conclude that the plane of oscillation of the pendulum at L must appear to turn about the vertical of that place with a velocity of rotation equal to that which the earth would have if it moved with the speed expressed by the component C B instead of that represented by the resultant O C. In other words, the velocity of the plane of oscillation is to that of the earth's rotation as C B to C A, but this ratio

$$\frac{BC}{AC} = \cos PCL = \sin LCE.$$

The velocity of rotation is therefore proportional to the sin : of latitude of the place at which the experiment is made.

The following demonstration is simpler than the preceding, and perhaps appears more evident to those who are not familiar with the theorems of mechanics:—

Let A (*fig.* 14) be the place of observation, and ϕ denote its latitude, draw A Z and the tangent A P which intersects the earth's axis in P, then the point A describes daily the parallel circle A A′ *a a′* while the vertical A Z traces the surface of a right cone. The pendulum swings in a vertical plane, and its intersection with the horizon indicates every moment the direction of the plane of oscillation. If the earth were at rest the angle between this line of intersection and the meridian would remain constant; it is now our object to find the changes of this angle, caused by rotation. To an observer standing at P the horizon at A will appear to revolve about a cone the apex of which is at P, while the vertical A Z through which the pendulum passes in each oscillation must always remain perpendicular to the revolving horizon. The point A retaining throughout the same distance from P will appear to a person standing at P as if it described

FIG. 14.

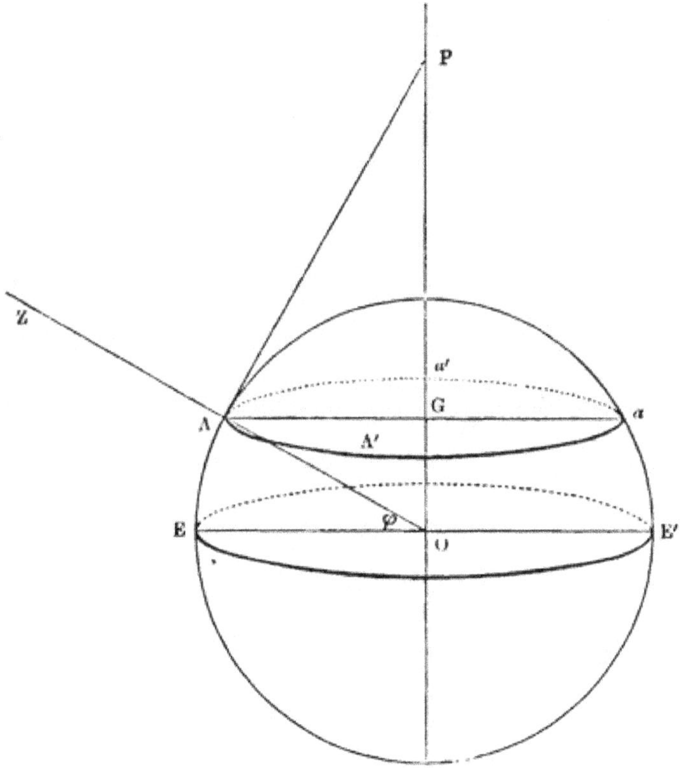

an arc A a with a radius P A. The length of the arc A a must be the same as that described on the parallel circle, but the corresponding angle at the centre P of this arc is smaller than that at G, the centre of the parallel circle. Therefore, while the point travels in twenty-four hours round the whole circumference A A′ *a a′* its apparent revolution round P will not be fully accomplished. The radii P A and P A′ represent in every position the meridian of the place. Let A B (*fig.* 15) be the direction of the plane

Fig. 15.

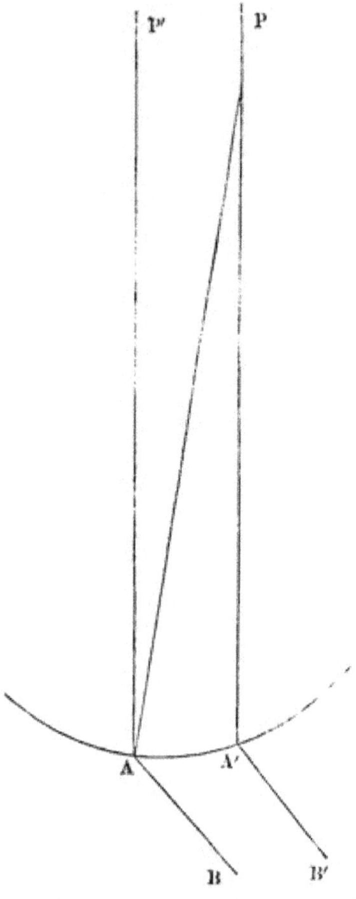

of oscillation in A, then we have for the direction of the same plane at the point A′ the section A′ B′ parallel to A B, the angle P A B has passed to B′, A′ P, i. e. the meridian has apparently moved from the plane of oscillation by the angle $APA' = \alpha$. Let ρ denote the angle of rotation between the positions GA and GA' and R the radius of the earth, then

$$AG = R\cos\phi \qquad AP = R\cot\phi$$
$$AA' \text{ in the circle about } G = \rho R\cos\phi$$
$$AA' \quad ,, \quad ,, \quad ,, \quad P = \alpha R\cot\phi$$

whence $\alpha = \rho\sin\phi$.

To this theoretical proof of the theorem relating to the change of the plane of oscillation we may add an experimental one due to the ingenuity of Professor Wheatstone. In the apparatus we use to show the principle of parallelism of the planes of oscillation of a pendulum suspended freely, we can only make the pendulum swing vertically with regard to the ring the plane of which produced is identical with the horizon, and, so long as we make gravity the moving force in our experiment, we cannot effect vibrations in planes which have any other inclination. Professor Wheatstone hit upon the expedient of substituting for the force of gravity that of elasticity. For this purpose he had a wooden semicircular arch A A′ constructed (see *fig.* 16), graduated in 180°, so supported at the centre C as to be movable round its vertical axis. This was intended to represent one of the earth's meridians, which, in the course of twenty-four hours, would complete a whole revolution; D D′ would therefore be a section of the equator. He then, in accordance with his adopted principle, applied a spiral spring C C′, one extremity of which he fixed on the diameter, so that its prolongation passed through the centre, the other was attached to a small clasp capable of sliding along the arch; by this means the chord C C′ could be brought to any position relatively to the equator D D′, the arc C′ D always representing the latitude of the place C′. If the chord be made to vibrate in a certain plane, and the arch is at the same time turned about its axis, the following remarkable phenomenon will become manifest, viz. that the planes of vibration will always remain parallel to the first, in whatever oblique position the chord may have been placed; but if the chord be placed at right angles,

FIG. 16.

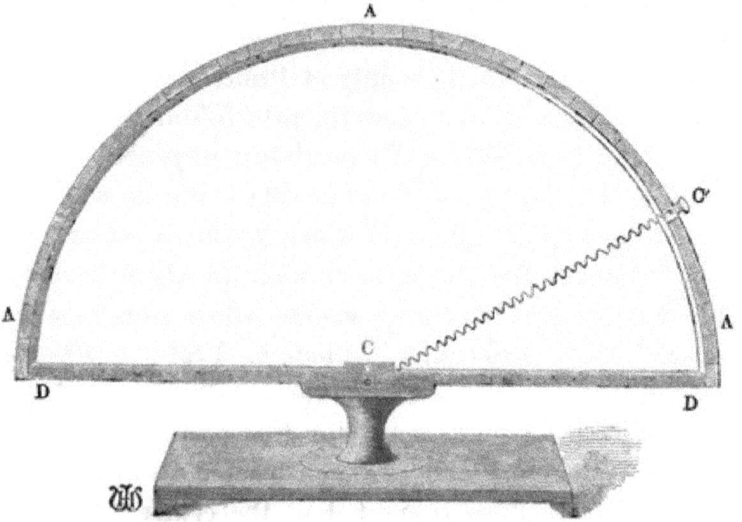

WHEATSTONE'S APPARATUS.

then the vibrations will always take place in one and the same plane, whence it follows that in this case the chord and the arch complete their revolutions in the same time, illustrating the motion of the free pendulum at the poles. In all other cases the revolution of the arch will be completed before that of the chord, and the ratio of velocity with which this takes place depends on the magnitude of the angle which the chord makes with the horizontal, or, in other words, on the degree of latitude. Let us suppose that the chord is placed at an angle of 30°, then it will be observed to make only half a revolution during the time that the arch completed the whole, and

$$\text{as } \sin 30° = \tfrac{1}{2},$$

the theorem is proved for this angle. A similar result would be obtained from any other angle, the fractional part of a chord's rotation during a complete revolution of the arch being expressed by the sine of the angle at which the chord is placed; but as the sines are in general irrational, we must be satisfied with an approximate verification.

Another apparatus for the same purpose was invented by Mr. Sylvestre. He wished to exhibit mechanically the simultaneous motions of the earth and different horizons as they really occur in nature. Let P K N E (*fig.* 17) be a

FIG. 17.

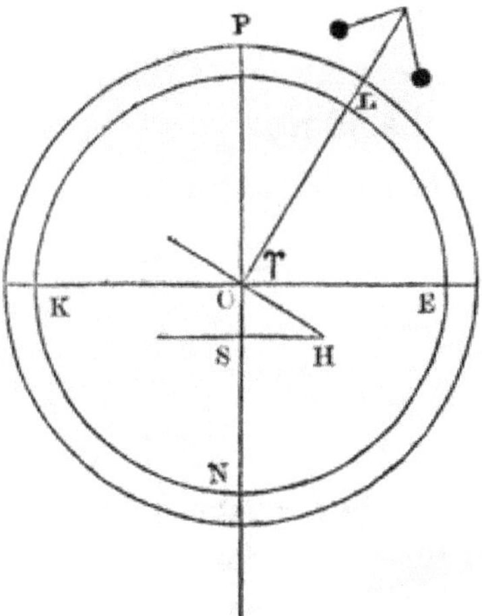

circular ring representing a meridian of the earth. $\angle$ is a point taken upon the circumference at any latitude, N P is the axis of the earth, O $\angle$ the vertical of the place of observation, and γ its angle of latitude. Let O R be the radius of a circular plane, the centre of which always coincides with that of the earth, and the axis of which is the line O∠ which passes through the line of observation. This circle, movable about its axis, and with a diameter which remains invariable after having been once chosen, is capable of taking different inclinations towards the plane of the equator according as the latitude changes. Let S R be the path upon the meridian of another circular plane the axis of which lies in that of the earth. This plane can only be raised or lowered along its axis, so that it will always be in contact with the circle of latitude, whatever the inclination of the latter may be. It is also obvious that the two planes O R and S R must always include an angle equal to the complement of latitude. According to the position of these two planes, which are dependent for motion on each other, the angular velocities of their circumferences in contact are in the inverse ratio of their radii. We have, therefore,

$$\text{ang. vel. } OR : \text{ang. vel. } SR = SR : OR$$

$$\text{but } SR = OR \sin \gamma$$

$$\therefore \text{ang. vel. } OR = \text{ang. vel. } SR \sin \gamma.$$

Hence, if we choose a certain place L, and then turn the ring about its axis, the latter will by its motion cause the vertical of the place to describe a cone, and the circle of latitude will also turn about its axis, so that when the ring has completed an entire revolution, the horizon of the place chosen will only have turned a fractional part, expressed by the sine of its latitude.

A similar and more practical instrument is shown in *fig.* 18; it was invented by a schoolmaster of Haverfordwest, and improved by Professor Wheatstone. On turning the stand round the ball revolves, but while it completes an entire revolution the arrow on the top of the axis will only describe a fractional part depending on the angle of inclination of the axis to the horizon. The graduated arc serves to set the axis at any angle required, and the screw fixes it in this position during the experiment.

FIG. 18.

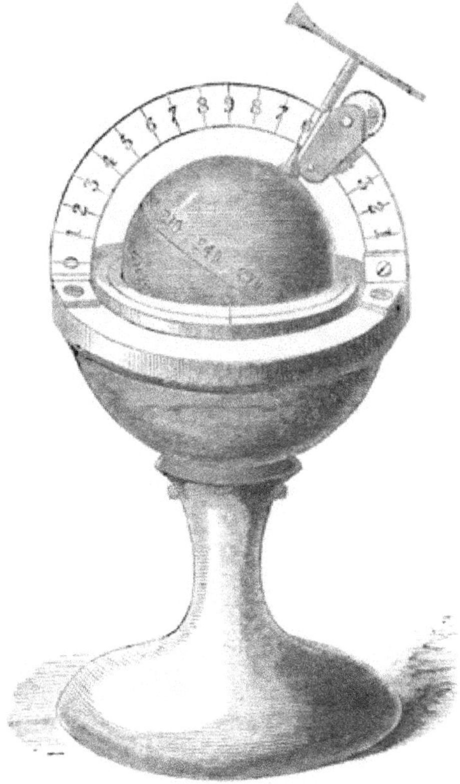

It now remains for us to compare with the results of observation the theory that the angular velocity of the plane of oscillation is equal to that of the earth's rotation multiplied into the sine of latitude. We may here prepare the reader for certain discrepancies which must occur in practice, and which are due to the following causes: the resistance of the air, the torsion of the suspending wire, and the lateral impulse. The disturbing influences of these forces we intend investigating theoretically in a subsequent chapter. We now proceed to state the results of actual observation.

CHAPTER VII.

EXPERIMENTS MADE WITH THE FREE PENDULUM TO PROVE THE ROTATION OF THE EARTH.

AFTER the well-known original experiments with the pendulum made by Foucault, those of General Dufour, who was assisted by Messrs. Wartmann and Marignac, are the next in importance. The pendulum which they used was 20 mètres in length, the bob, which was of lead, weighed 12 kilogrammes (about 26½ lbs.) and it was capable of swinging rather more than three hours, but it could only be observed during two hours and a half, for after that time the oscillations were too small. Four experiments were made in the meridian and four in the line perpendicular to it, commencing alternately from either extremity without any appreciable difference. Each experiment was continued till the pendulum had attained a deviation of 25°; the time was measured by a second-beat chronometer. The exact moment when the pointer fixed to the bob passed over one of the diameters, traced on the disc underneath, could be accurately ascertained so long as the amplitude of the oscillation was more than 0′ 70. The following is the mean result of four observations on each line:—

A deviation of 25° in the meridian was accomplished in	2,376 hours
A deviation of 25° in the meridian was accomplished in	2,110 „
Difference . . .	0,266 „

The difference is, therefore, about a $\frac{1}{4}$ of an hour. This proves that the velocity of deviation is not the same in the meridian as in the perpendicular, and the formula $a=\rho \sin \phi$ (ρ being the angular velocity of the earth, and ϕ the latitude of the place) is not rigorously true, for by it we obtain for a deviation of 25° at Paris 2,306 hours, a number which lies between the former two, but which is nearer to the first than to the second. This difference of time is so great that it cannot be attributed to an error in the experiment, but is owing to an acceleration in the line perpendicular to the meridian.

Mr. Morren repeated these experiments at Rennes and obtained similar results. The pendulum was 19·7 mètres in length, the bob weighed about 66½lbs., the amplitude of the first swing amounted to little more than three mètres. He also found that the deviation was greater when the oscillations were started in the perpendicular plane and less when in the meridian plane. He was well aware that these differences could not be due to currents of air or to the attractive force of the neighbouring walls, for the great mass of the pendulum rendered fallacious any such surmise, but he still could not arrive at the real disturbing power, which General Dufour conjectured to be the centrifugal force. In order to decide this point other experiments were necessary. On the 10th of February 1851, Mr. Bravais communicated to the French Academy the results of the experiments he made in the Great Hall of the Observatory with a pendulum ten mètres in length.

He employed two distinct methods — that of direct observations, and that of coincidences. By the direct method he observed the duration of from 900 to 1,200 levogyric oscillations of the pendulum, i. e. from west to east over the south, and then the deviation of a similar number of dextrogyric (from east to west) oscillations of the same pendulum; the distance from the point of suspension to the centre of the bob varied from 10187 to 10197 millimètres. The following table exhibits the observed deviations:—

	Levogyric oscillations	Dextrogyric oscillations	Difference
May 5	6,39887	6,39823	0,00064
" 11	6,39925	6,39849	0,00076
" 16 (A)	6,39815	6,39751	0,00064
" 18 (A)	6,39959	6,39863	0,00096
" 16 (B)	6,40106	6,40032	0,00074
" 18 (B)	6,40116	6,40044	0,00072
	Mean difference		0,00074

The difference furnished by theory is equal to 0,000716, which agrees satisfactorily with the above result.

The other method offers still more accurate results. Two pendulums of different lengths, but similar in other respects, are placed in the plane of the meridian at a distance of 7 decimètres from each other. The space between the point of suspension and the centre of the bob was 10216 millimètres for the longer pendulum, and 10115 millimètres for the shorter. The levogyric and dextrogyric oscillations were communicated to the two pendulums respectively at the same time, and the moment of passing the meridian was observed by means of a telescope, the optical axis of which intersected the two threads when at rest. The number of oscillations between two successive coincidences of the two pendulums was noted down; let them be $n\, n'$, then $n' = n + 1$ where n' refers to the short pendulum. The series of observations being finished, the pendulums were stopped and afterwards set in motion with reversed directions. The coincidence was again observed, and the two new numbers determined, N and $N' = N + 1$ denoting by λ the latitude of the place, τ the durations of a sidereal day, t the duration of the conical oscillation of the long pendulum uninfluenced by the effect of terrestrial rotation, we must obtain, if the theory be exact,

$$\frac{\sin \lambda}{\tau} = \frac{1}{t + t'} \left\{ \frac{1}{n + n'} - \frac{1}{N + N'} \right\}$$

The experiment of the 25th of May gives

$n = 207{\cdot}86$ $N = 217.82$

$n' = 208{\cdot}86$ $N' = 218{\cdot}82$

The experiment of the 10th of June gives

$n = 206{\cdot}31$ $N = 215{\cdot}96$

$n' = 207{\cdot}31$ $N' = 216{\cdot}96$

After making the necessary corrections, we derive from these the difference of duration between the levogyric and dextrogyric oscillations:

Difference found May 25th. . .	0·000725
,, ,, June 10th. . .	0·000710

Whence it follows, that for a pendulum of 10 mètres which turns from west to east, the angular velocity is retarded at Paris by 11.4 seconds of space for each second of time, and that it is increased by the same quantity when the rotation takes place from east to west.

This inequality in the duration of dextrogyric and levogyric oscillations of the free pendulum was confirmed by experiments made in Rio de Janeiro (in 22° 54′ south latitude) during the months of September and October 1851 by Mr. D'Oliveira. The following is an account of his experiments:—

Mass of the pendulum employed.

A hollow sphere 24lbs. in weight furnished with a pointer.

Suspension of the pendulum.

An untwisted hempen thread fastened to the ceiling, and having at its other extremity a piece of iron fitting into the hollow sphere, so adjusted that the extremity of the pointer corresponded with the point of suspension, the length of the pendulum being 4,365 mètres.

Arcs described by the oscillations of the pendulum.

The first experiments were made by imparting to the pendulum a swing of 5° 14′ 44″, but in the later ones the arcs were 7° 51′ 41″. The arcs corresponded to the length of the oscillations measured by their tangents, which amounted in the first to 4 decimètres, and in the second to 6 decimètres.

Trajectory described by the pendulum.

The trajectory formed almost always in a double oscillation a very elongated ellipse, the minor axis being hardly large enough to show the deviation of the pendulum by the tracings in fine sand made by its pointer. The square frame containing the sand had the cardinal points marked on it, and so placed under the pendulum that the intersection of two diagonals corresponded with the extremity of the pointer.

The pendulum was put in motion first in the direction

of the meridian, and started alternately from the north and south sides, and afterwards the experiments were repeated in the direction of the parallel circle. Similar experiments were also made in several directions intermediate to these, two south-west and south-east. The results obtained from those experiments, which were considered as accurately carried out, are the following:—

1st. The motions of the pendulum describing ellipses in the meridian or parallel circle always took place in opposite directions, that is, when the ellipse in the meridian was traced from right to left (like the rotation of the earth looking towards the south pole) the pendulum moved from left to right in an ellipse described in the direction of the Parallel. A reduction in the thickness of the thread of suspension lessened the length of the minor axis.

2nd. Lifting the pendulum three decimètres from the vertical, in order to make it describe the greatest arc, and allowing it to swing during 30° first in the direction of the meridian, and afterwards in the direction of the parallel circle, the deviation was found in the first case to be 5° 13′ towards the east, and in the other case 5° 12′ towards the south.

3rd. Measuring the lengths of the major axis of the ellipses by means of the tracings on the sand, they were found to be $\frac{356}{600}$ mètres and $\frac{349}{600}$ mètres respectively with regard to the meridian and Parallel.

4th. Moving the pendulum in various directions, a line was found along which the trajectory had no tendency to move either to the right or left, and at the same time the pendulum continued to move without deviation, so that this line marked out the position of an invariable plane of oscillation. This plane made with the Parallel an angle of 11° 18′ 40″, or very nearly half the latitude of the place of observation.

5th. The motion of the pendulum made in two directions midway between the meridian and Parallel, showed that the deviation of the south-west portion was much slower than in the meridian, while that in the south-east was quicker; but all these deviations were in the same direction, namely, in that of the rotation of the earth.

Before proceeding to give an account of our own observations, we will give the opinion of a great authority on an experiment made in Turin. Plana, in the 'Memor. dell Acc. di Torino,' xiii. 55, 56, says that the time of an entire revolution of the plane of oscillation was found to be 34 hours 49 minutes instead of 33 hours 48 minutes, according to theory. To clear up this difference the experiment was repeated, but with a worse result, for this time the whole revolution occupied 40 hours. Plana justly attributed this to the defects of Foucault's mode of suspension, which it will be remembered consists in squeezing one end of the thread between the metallic plates screwed to the ceiling (see *fig.* 9). The disadvantages of this mode of suspension are the bending of the thread at its upper extremity, and its twisting so as to cause a motion of the pendulum round its axis; the former causes the amplitude of the swing to diminish, the latter tends to change the azimuth of the plane of oscillation. The disturbances produced by the torsion may be calculated provided the thread is homogenous throughout, and its section a perfect circle. It is impossible to preserve the cylindrical form of the thread at its upper extremity owing to the compression of the two plates: hence changes will occur which vary with the different azimuths, and are consequently beyond calculation. It was therefore evident that another mode of suspension was required which should give these researches a higher degree of accuracy. Hansen suggested the following method:—Instead of a fine thread, a thick wire should be employed, so that torsion would be impossible; the bob is so fastened that its centre of gravity lies in the prolongation of the axis of the wire, the upper end is curved, and teminates in a spherical segment of hardened steel, which, placed on a horizontal plate of the same material, forms the suspension (see *fig.* 19).

FIG. 19.

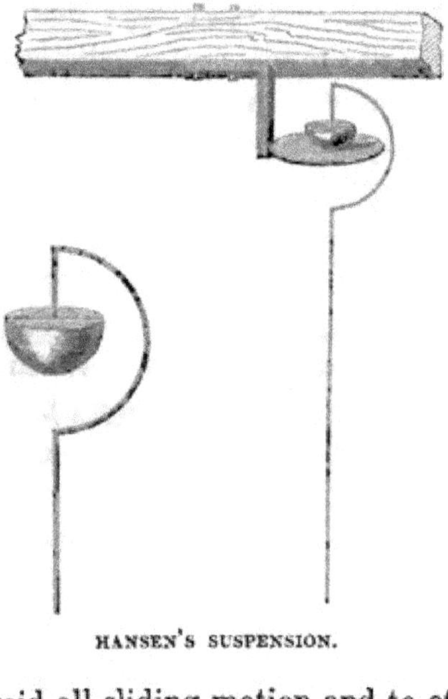

HANSEN'S SUSPENSION.

In order to avoid all sliding motion and to effect a rolling one, during the swing of the pendulum, the segment and plate are ground but not polished, by which friction is greatly diminished. Though the theory of this arrangement has been very ingeniously developed by the learned inventor, yet we are not aware that as yet any practical trial of it has taken place. We ourselves, in conducting some experiments, made use of a suspension similar to that proposed by Hansen, with one essential difference, viz. a cone was substituted for the spherical segment working freely in a corresponding socket (see *fig.* 20). Another mo-

Fig. 20.

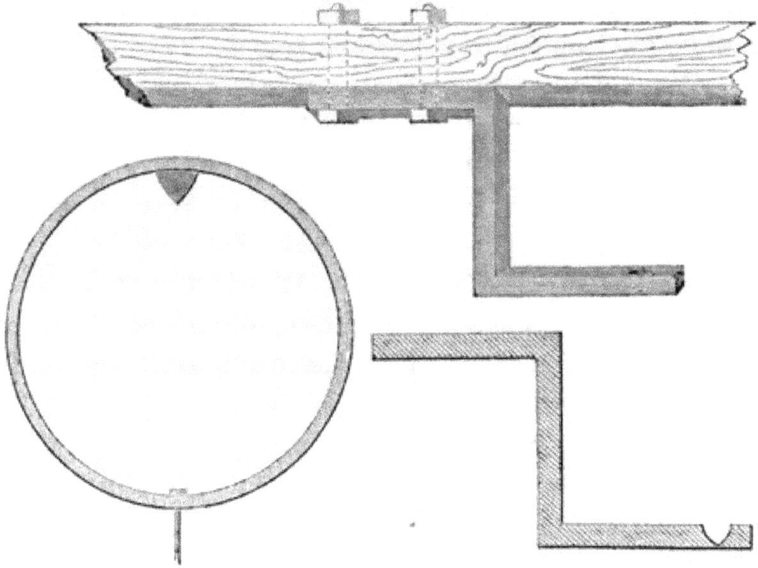

DETTMANN'S SUSPENSION.

dification was made in the mode of liberating the pendulum; in all previous experiments the bob was drawn up to the required angle and maintained in that position by a thread, which was either cut or burnt after perfect rest was obtained. Now, it will be easily understood that by this method the particles of the thread cannot be severed simultaneously, so that there is a slight jerk before complete liberation. To obviate this, I fixed on the graduated board underneath the pendulum, at the proper distance, two small perpendicular posts pierced at the top by two horizontal screws, the points of which were accurately ground and capable of coming together; one end of the loop destined to hold the pendulum was then compressed between them, the pendulum was then raised till the pointer could be looped in; after a short interval, to allow the instrument to become quiescent, a turn of one of these fine-threaded screws set the pendulum free instantaneously, while the loop fell to the ground (see *fig.* 21).

The following is a short account of the apparatus used, and the experiments I made, in the Hall of King's College, London, in the year 1859: —

The bob was a truly-turned ball of brass, weighing 40lbs., the suspending medium was a thick steel wire, the length of the pendulum was 17 feet 9 inches. The amplitude of the first oscillation, as measured by its tangent, amounted to 6° 42′ 22″, and during the time of the experiment, which was about half an hour, the arcs were not much diminished. As I had to demonstrate to a large number of spectators I encountered considerable difficulty in rendering the small deviations of the plane of oscillation visible to all. I accomplished this in the three following ways: —

1st. I placed on the first two lines of division (counting from the meridian in which the pendulum commenced its swing) two small wooden cones, so that the diameters of their bases coincided with the line that marked the degree; their weight was regulated so that when the plane of oscillation deviated sufficiently to pass through the points of their azimuth, they were struck by the pointer

Fig. 21.

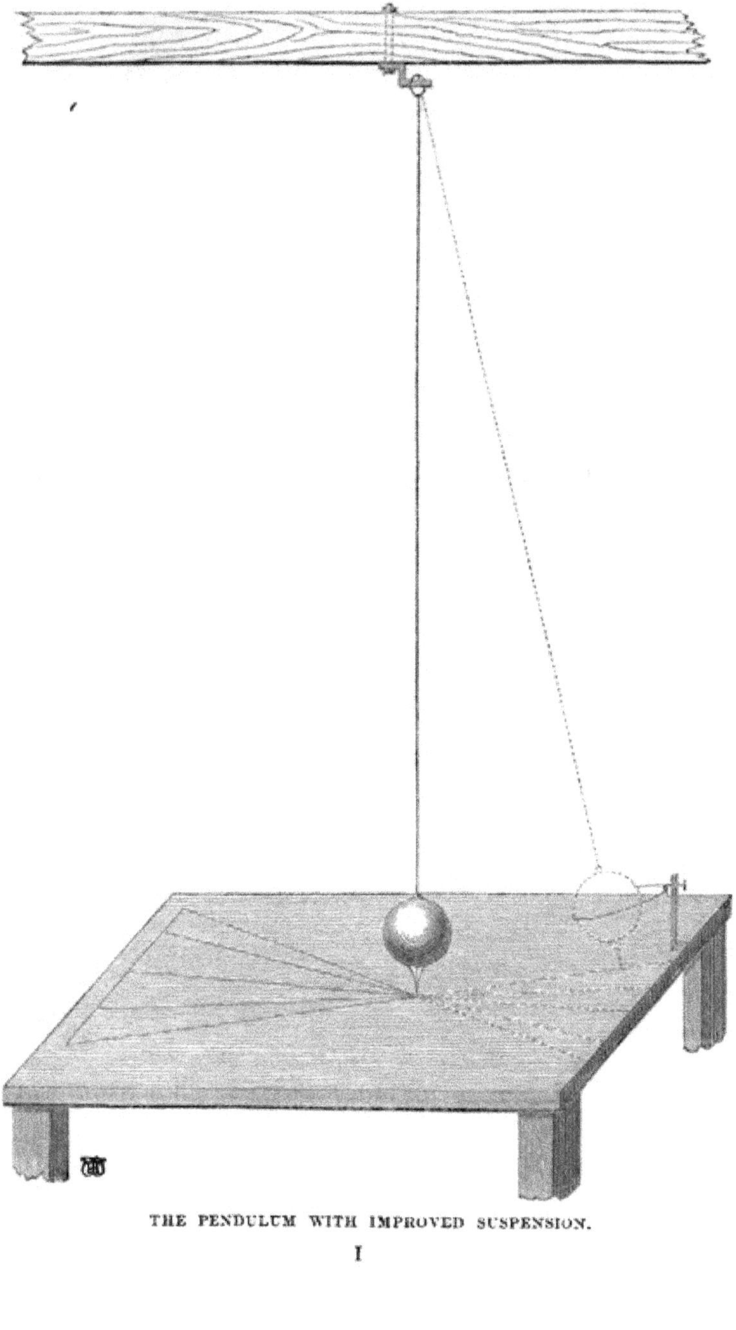

THE PENDULUM WITH IMPROVED SUSPENSION.

Fig. 22.

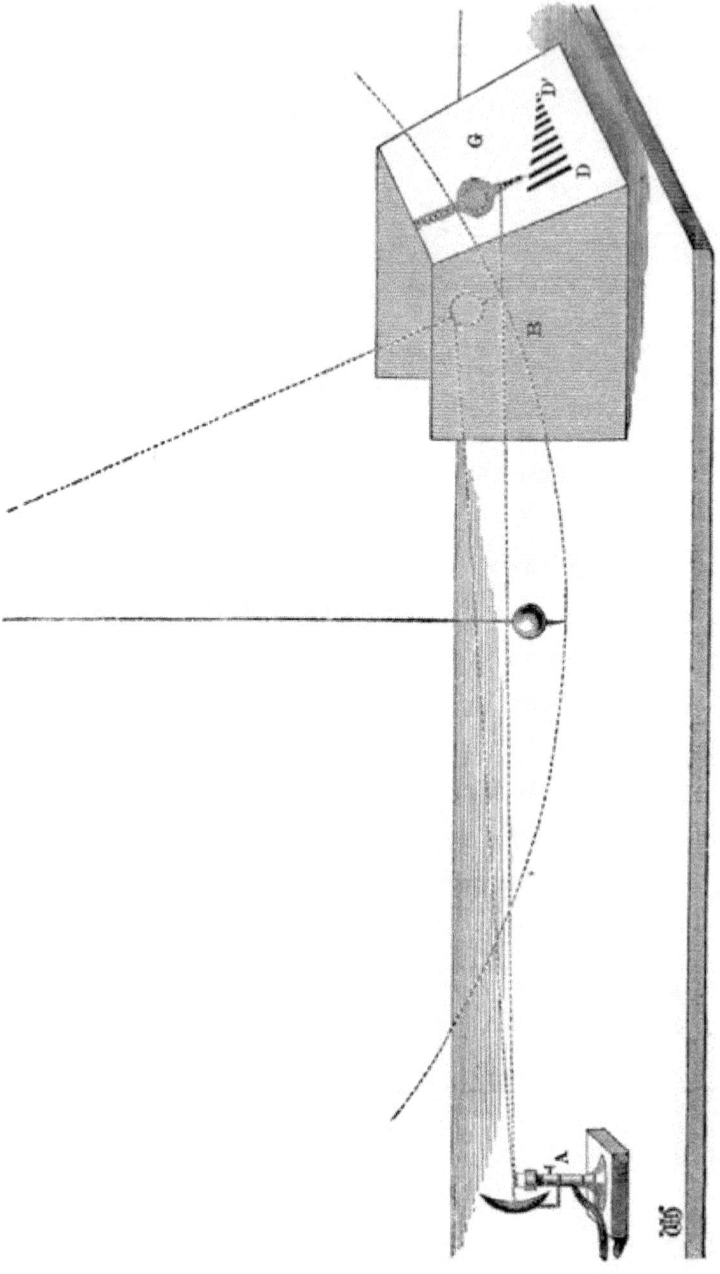

WORMS' APPARATUS FOR SHOWING THE MOTION OF THE PENDULUM.

and knocked down. By this means everyone present was enabled to observe the time taken for each degree of deviation.

2nd. At the fourth degree from the meridian line I placed a small cannon loaded with gun cotton, across the touch-hole of which was a fine platinum wire in connection with one of the poles of a galvanic battery, the metallic suspension being connected with the other pole. The object of this arrangement was that, when the deviation amounted to four degrees, the pointer should come in contact with the platinum wire, thus completing the circuit and firing the cannon. This effect took place after the proper time furnished by theory.

3rd. The third method will be best understood by reference to the figure. (*Fig.* 22.) All lights in the apartment having been lowered, the lamp A was lighted. B represents a rectangular box open at the top and back. The front G being of ground glass and slightly slanted, D D′ are divisions corresponding to the degrees of deviation. A well-defined shadow of the bob and pointer was by this arrangement projected on the glass, and the gradually increasing deviation could throughout its progress be observed with the greatest ease by every person in the room, and the moments of passage through the points of division noted down. The results on the whole were most satisfactory, owing mainly to the accurate suspension and mode of liberation, for, as far as ocular observation went, no elliptical swing took place.

Although the number of pendulum experiments which we have described in this chapter is but small, yet we venture to believe that they afford sufficient evidence of the earth's rotation.

CHAPTER VIII.

THE GYROSCOPE.

IN the autumn of 1852 Mr. Foucault presented the Academy of Sciences of Paris with another experimental proof of the earth's rotation, which was founded upon the fixity of the plane of rotation of a body suspended freely and revolving about one of its principal axes. The instrument he used he called the 'Gyroscope;' many modifications have since been made, and indeed, many years before Foucault's invention, Bohnenberger, a German astronomer, constructed a similar apparatus, which he employed to show the precession of the equinoxes, and which, by the suggestion of Laplace, was introduced in the Ecole Polytechnique for the purpose of explaining this phenomenon. In order fully to understand Foucault's application of the gyroscope, it is necessary to have a clear conception of the laws of rotation, which form a very difficult part of mechanical science, and as the modified form now so commonly used demonstrates them practically, we will first give a description of it.

It consists of a heavy solid disk D (*fig.* 23), turned with extreme accuracy on an axis A A′, which terminates in points P P, the centre of gravity of the disk and axis being exactly at the middle of the line joining the extremities of the axis. These points are inserted into suitable sockets E E, on opposite sides of a light but strong brass ring R R. At the extremities of the diameter of this ring, perpendicular to the

Fig. 23.

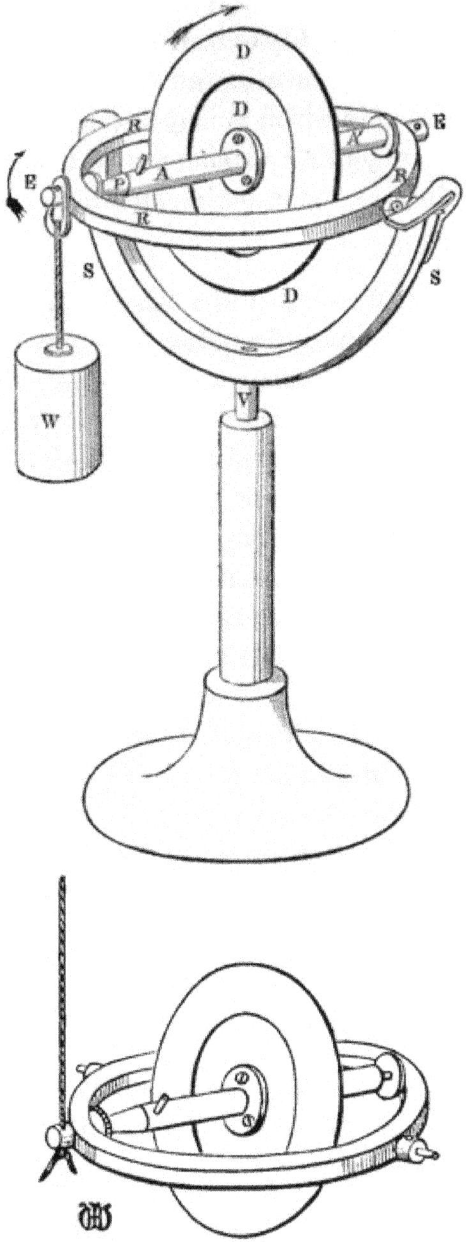

THE GYROSCOPE.

axis, it is inserted in the same manner, into another ring or semicircle S S, which by means of a vertical axis V secures the perfect mobility of the inner ring, and thus the central heavy disk can turn in any direction, or remain at rest in any position in which it may be placed. The disk is made to rotate by pulling a string wound round the axis.

The following are the experiments which can be made with this instrument: —

First experiment — When the wheel is in rapid rotation, detach the inner ring from the outer, lifting it carefully together with the revolving solid which it supports; then, holding the ring firmly in one or both hands, attempt to change the position of the body. It will be found that when the solid is rapidly revolving, the difficulty of displacement is very great, and the displacement which is effected is perpendicular to the direction of the force you apply. This is a phenomenon in itself remarkable, and the explanation of it is the key to most of the other experiments with the instrument, and generally to the peculiarities of rotation, which it is intended to illustrate. This may be expressed in the form of a theorem as follows. When a disk is thrown into very rapid rotation round an axis, and another power tends to turn it round another axis perpendicular to the first, then the result will be a rotation round a third, perpendicular to the plane containing the other two; i. e. the power applied may be said to act 'tangentially.'

Second experiment — Suspend the apparatus which you had removed for the last experiment by a string attached at the outer part of the ring at the extremity of the axis of the revolving solid. The ring being then placed in a horizontal position, the string held in the hand, and the apparatus left to pursue its own course, it will be seen, contrary to expectation, that the apparatus instead of falling straight down, as it would do if not rotating, retains its horizontal position, and at the same time marches round the string with a slow uniform movement. The string remains during this motion in a vertical position. The external forces impressed on the solid in this case are the weight of the wheel acting downwards at its centre of gravity, and the force of the string, which is equal to the weight of the wheel, acting upwards at the point of suspension. This pair of forces would evidently produce a rotation about a horizontal line in the equator of the wheel, and we see that their effect is, like the forces applied in the first experiment, to make the axis of the wheel deviate towards this line.

Third experiment — Let the apparatus be replaced in its stand. Having set the wheel rotating, hang the smaller weight furnished with the instrument at the point where the string was applied in the preceding experiment. Then, if the fixed vertical circle be set free, this weight with the whole apparatus will be carried round with a slow uniform rotation in a horizontal plane. The explanation of this will evidently be the same as that of the last experiment.

Fourth experiment — While things are in the condition just described, hang the heavier weight at the other end of the axis; the motion will then be reversed.

Fifth experiment — While the wheel is rapidly rotating, and the vertical semicircle perfectly free, hang the weight as in the third experiment, and while the apparatus is moving round suddenly fix the vertical semicircle. The moment this is done the precessional motion ceases, and the weight falls as it would have done from the action of gravity, if there had been no rotation. It appears, then, from this experiment, that the supporting of the weight contrary to the natural effort of gravity is only effected by means of the rotation of the ring, when the accompanying precessional motion is not prevented.

Sixth experiment— In the last experiment it appeared that when the precessional motion is prevented, the force applied to the axis has the same effect whether the wheel be in rotation or not. To illustrate this further, remove the weight, and set the wheel in rotation. It will then be found that the wheel may be turned completely over, as if it were not in rotation.

If the gyroscope were made to revolve with great rapidity, its axis would, if perfectly unrestrained, retain a constant absolute direction, and therefore perform an apparent revolution about the direction of the earth's axis, just as the star to which it points. If, however, the axis be restrained, so that it can only move in a given plane, it will oscillate about that line in the plane, which is nearest to the earth's axis. These oscillations are isochronous, and will be more frequent the greater the velocity of rotation.

This principle may be thus enunciated in the words of Foucault: —

When a body turns upon a principal axis, and a force or system of forces tends to produce another rotation not parallel to the former, then the resulting effect is a displacement of the axis of original rotation, which directs itself in the way most favourable to the parallelism of the two rotations. The new axis of rotation may be either fixed or movable, and dependent on the position of the body; when it is fixed, the turning body tends to a definite position, and when movable with the axis of the body, the latter changes its direction continually, without ever arriving at a state of equilibrium. In order to elucidate these remarkable facts of rotary motion, we cannot do better than introduce to the reader the gyroscope, invented by Mr. Fessel of Cologne (see *fig.* 24), which is perhaps

Fig. 24.

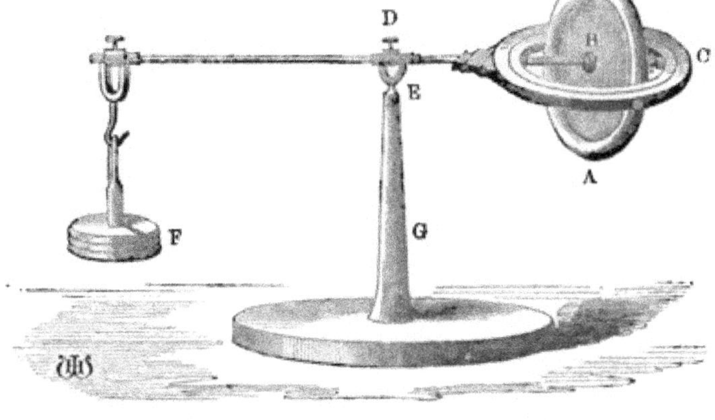

FESSEL'S GYROSCOPE MODIFIED BY WHEATSTONE.

the best adapted for the exhibition and explanation of them, especially in the modified form given by Professor Wheatstone. A beam is capable of moving round a horizontal axis, which is itself movable about a vertical one, so that it may be turned in any direction about a fixed point. At one end of the beam a horizontal ring is fixed, which carries a heavy disk, the axis of which is in a line with the beam. At the opposite extremity a shifting weight is attached, by means of which the equilibrium may be established or disturbed at will. If the instrument be put in equilibrium, and the disk be made to revolve rapidly by means of a string wound round its axis, then the beam has no tendency to displace itself in any direction; but on moving the weight towards its centre, and causing the disk to preponderate, then, if the latter rotate from right to left, the former will turn about its vertical axis in the same direction, and vice versâ. If the weight be allowed to preponderate, contrary effects will take place. The velocity of rotation round the vertical axis increases in proportion to the amount of disturbance of the equilibrium, but notwithstanding the increased or diminished action of gravity on the disk, its axis of rotation always preserves that inclination to the vertical axis at which it was originally placed. When the equilibrium is disturbed while the disk is at rest, gravity acts so as to turn the beam round a horizontal axis, that is, about an axis perpendicular to the plane which contains the axes of the two original rotations.

A similar composition of forces takes place by impressing on the beam in equilibrium a motion round the vertical axis. If the disk rotates at the same time from right to left, then the end of the beam conveying the latter will ascend, provided that the pressure tending to produce motion round the vertical axis acts in the same direction, but if it act in the opposite, it will cause the disk to descend; hence, as in the preceding case, the beam is constrained to move about a horizontal axis perpendicular to the plane which contains the other two. The beam ascends and descends in a like manner whenever the motion round the vertical axis is accelerated or retarded. If the disk rotate from right to left with its own weight predominating, the rotation round the vertical is from left to right. Accelerating it causes the disk to descend, and retarding, to ascend.

Let the ring carrying the disk be removed, and the tightening screw unfastened, then the axis of the rotating disk is free to move in a plane; and if the outer is constrained to move about the line in this plane, the movable axis will place itself so as to coincide with that line; but if the fixed axis be in a different plane, the movable will assume that position which approaches most to the former. This phenomena may be thus generally explained.

Suppose the apparatus be in the position indicated in the figure, let the axis D be held horizontal, the weight at F being removed, and let the disk be made to rotate so that its front A has an ascending motion. This rotation communicates to all parts of the disk tangential velocities, which may be resolved into their horizontal and vertical components. Let the ring now be liberated, so that it can be acted upon by the force of gravitation. The next effect will be that the axis of rotation sinks down slightly, and the disk in consequence takes an oblique position; the horizontal components of the former motion are not changed by it with respect to the position of the disk, but the vertical are: they leave its plane, and are in front when ascending towards the left, and behind when descending towards the right. A lateral force is consequently exerted upon the disk, which draws its front towards the left, and its back towards the right. Both actions supporting each other effect a rotation of the whole apparatus round the vertical axis E, from left to right. The striking phenomenon of the pressure on the axis of a freely rotating body acting in a direction perpendicular to the plane of rotation is therefore easily explained, but this effect does not terminate the process of actions which modify the motion, for as soon as the disk begins to turn about the vertical axis E, the horizontal components leave the plane of rotation, and exercise a force upon the disk which draws its lower part to the right, and its upper to the left. The resultant of these two directions is an ascending motion of the disk about the horizontal axis D; that is, contrary to the direction im-

FIG. 25.

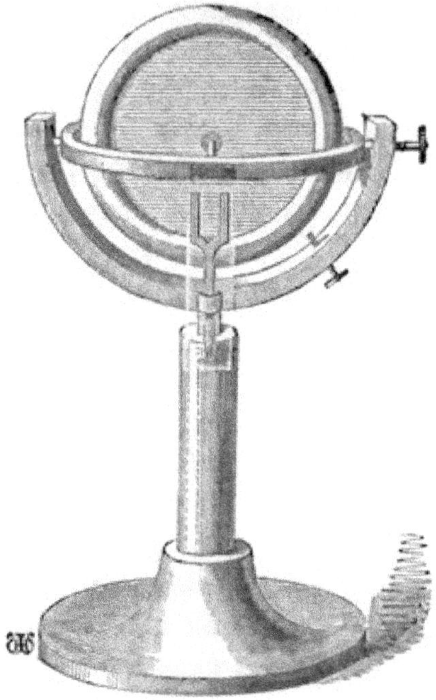

WHEATSTONE'S MAGNETIC GYROSCOPE.

parted to the system by gravity, yet equal to it in power. Hence, by putting weights at F, the horizontal motion may be stopped, or changed into one from left to right. Other forces may be substituted for weights to act upon the ring of the gyroscope, which tend to produce similar effects. This has been practically carried out by Professor Wheatstone in his magnetic gyroscope. (See *fig.* 25.) In this instrument a portion of the ring consists of magnetised iron, immediately underneath which, at a slight distance, is a small horse-shoe magnet, movable on a vertical axis; by this means its poles can be reversed, while those of the upper magnet remain fixed. By this arrangement a repelling or attracting force can be applied as a substitute for putting on or removing weights.

Another apparatus worthy of mention was invented by Professor Magnus of Berlin. It consists of two disks (see *fig.* 26) A B and C D, which can be put in rapid rotation, with a nearly equal amount of velocity, by means of the handle E F; these are supported by arms connected by the rod *m n*, which can turn about the horizontal axis *q r*. The apparatus rests on the vertical axis *v w*, round which the whole moves easily. The screw *z* serves to arrest the motion about the horizontal, and the piece of metal *t u* to arrest, accelerate, or retard that about the vertical. The following phenomena exhibit the results of combined rotation: —

1. If both disks rotate in the same direction, and no weights or equal ones are suspended at *m n*, then the axis *m n* remains in its original position.

2. If the weights suspended on both sides of *m n* are of different magnitudes, then the whole apparatus will turn about the vertical axis in one or other direction, according to the side on which the weight predominates, whereby the position of the axis *m n* is unappreciably altered, even though the overweight be very great.

3. If the motion round the vertical axis be accelerated by means of the small rod *t u*, then the loaded disk will rise, and vice versâ.

Fig. 26.

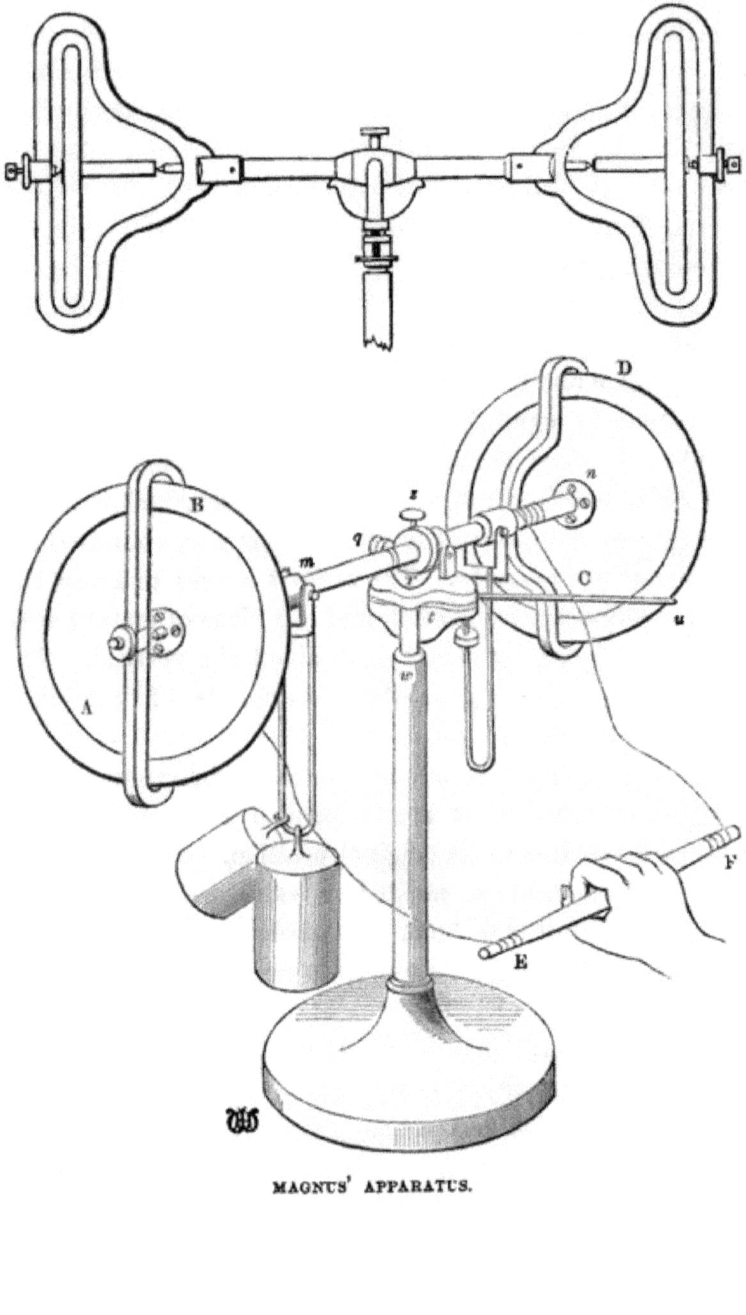

MAGNUS' APPARATUS.

4. If the screw z be tightened while the apparatus is in motion, then the rotation about the vertical axis ceases suddenly, and again commences on its being loosened.

5. If the disks rotate with equal velocities in opposite directions, then the side which is the heavier loaded will sink.

6. If the disks are made to rotate with opposite directions and equal velocities, after having been placed at unequal distances from the centre, the whole apparatus will be found still extremely sensitive. Whence we conclude that the distance at which a revolving body is placed from the vertical axis has little influence on the motion of the instrument. The explanation of these phenomena rests on the same principle as that of Fessel's gyroscope.

The instruments we have described, although they illustrate the general phenomena of rotation, are not capable of rendering apparent the diurnal motion of the earth: for this purpose, not only greater accuracy, but additional arrangement in the contruction, are required.

The principal parts of M. Foucault's gyroscope are, as in the other instruments, the heavy disk and ring; but the steel axis of the former is provided with a toothed wheel, and from the ring two knife edges project in a direction perpendicular to it. By this means the apparatus, with its ring in a horizontal position, can be placed in connection with a combination of multiplying wheels, in order to impart to it a very rapid rotation. This having been effected, it is placed with its knife edges upon a vertical ring, which is suspended by a thread without torsion, and rests on a pointed pivot. (See *fig.* 27.) The small screws about the ring, one pair of which is movable in a horizontal, the other in a vertical direction, serve to adjust the apparatus, so that the centre

Fig. 27.

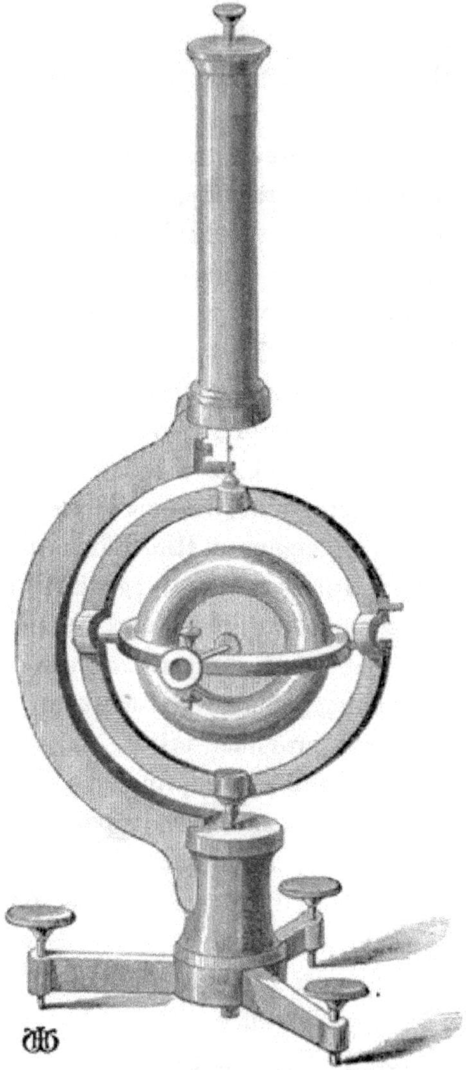

FOUCAULT'S GYROSCOPE FOR SHOWING THE ROTATION OF THE EARTH.

of gravity may lie in the prolongation of the suspending thread. If this is well carried out, the force of gravity can have no influence either on the rotation of the disk, or upon the entire system. The plane of rotation of the disk remains fixed, therefore, in the original position, not participating in the diurnal motion of the globe. The relative displacement arising from it may be observed by a microscope placed near the apparatus. Foucault described the phenomenon in the 'Journal des Débats' of September 22, 1852, in the following words:—

'The axis of a whirling body being fixed in absolute space, appears when examined under the microscope to retrograde from east to west, and travels in the same continuous manner upon the glass of the instrument as the image of the celestial bodies in the focus of the astronomical telescope. I have furthermore established by experiment a singular property of rotating bodies, which I found à priori by pure reasoning—I mean the force of orientation, which tends to place the axis of the body parallel to that of the earth, and at the same time to place the two rotations in the same direction. This force of orientation shows itself always when the axis of the whirling body is constrained in a fixed plane, but can move freely within it. The principle which has guided me in my researches on the subject may be enunciated as follows. When a body turns upon a principal axis, and a force or a system of forces tends to produce another rotation not parallel to the former, then the resulting effect is a displacement of the axis of the original motion, which places itself in the most favourable direction for the parallelism of the two rotations. The new axis of rotation may be either fixed or movable, and dependent on the position of the body.

When fixed, the turning body tends to a definite position, which is in accordance with the motion of the earth ; when movable with the axis of the body, the latter changes its direction continually, without ever arriving at a state of equilibrium.

These experiments of M. Foucault to render the rotation of the earth visible have been made the object of mathematical investigation by Mr. Quet, but as we purpose giving another analysis of it in the second part, we will content ourselves with stating here the principal results this eminent philosopher arrived at. He supposes the centre of gravity of the turning body to be fixed, and its axis movable only in a certain plane, which is also fixed, termed the directing plane.

The following are the various modes in which the axis influenced by the diurnal motion of the earth oscillates.

1. When the directing plane is horizontal the axis of the body can only be in equilibrium when in the meridian ; this equilibrium is stable if the rotation of the body projected upon the terrestrial equator lies in the same direction as that of the earth, otherwise it is unstable.

2. When the directing plane lies in the meridian the axis of the body can only be in equilibrium when in a position parallel to the terrestrial axis.

3. The axis of the rotating solid will assume the same position as in the preceding case when the plane itself lies in a position parallel to the terrestrial axis.

4. When the directing plane is perpendicular to the earth's axis then that of the body is in indifferent equilibrium.

5. In general, whatever be the direction of the plane, we obtain by projecting the earth's axis upon it the direction of equilibrium of the axis of the rotating body.

6. When the axis of the body lies without its line of equilibrium, it oscillates in a directing plane about its stable position. The oscillations, whether great or small, obey the same laws as those of the pendulum.

7. If the axis of the body is made to oscillate alternately in the meridian and in the horizontal plane, we find for the same velocity of rotation that the vibrations in the former are quicker than in the latter.

8. The duration of the oscillation in the meridian may serve to determine the time of the earth's rotation.

9. Comparing the squares of the numbers of the oscillations which take place respectively in the horizontal plane and in the meridian, during equal intervals, and with the same velocity of rotation, their ratio will give the cosine of the latitude of the place of observation. Hence, supposing an instrument to be accurately constructed on the principles laid down by the theory, an observer could determine by it, without seeing the heavens, the direction in which the stars seem to revolve, and the duration of the revolution, which would afford a most convincing proof of the earth's diurnal motion.

We cannot conclude this chapter without laying before the reader an objection to the primary hypothesis made by Mr. Foucault, with regard to his gyroscope, viz. that its only connection with the earth is through its centre of gravity. This cannot be exactly true, for neglecting even the torsion of the thread by which the ring is suspended, the friction of the pivots is sufficient to communicate the earth's motion to the gyrating body. This state of things does not exist in Bohnenberger's original apparatus, which we will now describe, and explain its application.

The principal part of this instrument, see fig. 28, is a

massive globe * capable of turning in all directions about its centre of gravity, which rests in a ring, which itself is suspended within two other rings. If the ball be set to

Fig. 28.

BOHNENBERGER'S GYROSCOPE.

rotate about a horizontal axis, then the diurnal motion of the earth, or in our latitudes, the component part of it, will tend to move it about the vertical. The axis will, instead of a horizontal motion, receive one in a plane perpendicular to the horizon, thus furnishing us with a plane invariable in the azimuth, by means of which we can easily see whether the earth turns or not. Strictly speaking, this plane is only indicated by the extremities of the axis, but the apparatus offers a more substantial one, in the shape of the middle ring, which, like the sphere, cannot move in the azimuth; its stability is most

* The substitution of a spheroid in the place of a globe is not advisable, as it leads to the erroneous opinion that the phenomena observed depend on the flatness of the figure.

striking, for we may turn the instrument in whatever way we please, it will always return to its original position. Therefore, inasmuch as the instrument is not constrained by the rotation of its support, it affords a means of verifying that of the earth; it must, however, be borne in mind that our assumption of a sole tendency to move round the vertical is only true at the poles; at any other latitude, the component horizontal of the terrestrial rotation interferes, which complicates the phenomena. We will now examine this force, and show how, by elimination, we may always produce the same results as at the poles. Let Z P H represent a meridian, λ the latitude of the place Z, n the angular velocity of the earth on its axis, C P, then by Euler's theorem we have $n \sin \lambda$ and $n \cos \lambda$ for the angular velocities about C Z, and the horizontal C H. Supposing Bohnenberger's apparatus to be at Z, its rotation about the vertical is obvious, since the latter passes through the centre. The rotation about the horizontal, C H, may be resolved into a translation in a straight line, and into a rotation; during the translation the apparatus remains parallel to itself, turning at the same time with an angular velocity $n \cos \lambda$. The motion of translation having no influence in this case, a "couple" is generated, the axis of which lies in the meridian and the forces in the first vertical; the axis will, therefore, move in the azimuth. Such motion, however slow, would render doubtful the observations made to prove the rotation of the earth, but this difficulty may be obviated in the following way. The two components of terrestrial rotation, acting separately, impart to the apparatus the angular velocities $n \sin \lambda$ and $n \cos \lambda$; therefore acting together they give it the velocity n, about an axis parallel to that of the earth. Let us now place the general axis of the instrument in this direction, then the vertical component of terrestrial rotation is replaced by the total rotation, and no force exists which tends to give it a motion in the azimuth. The earth's rotation communicates to the little globe a very slow motion in the meridian in which it was placed, whereas the apparent motion in right ascension is much quicker, being equal to that of the earth, so that the middle ring will appear to make a complete revolution in twenty-four hours as at the poles.

We have thus pointed out the arrangements by which Bohnenberger's apparatus may in all latitudes be used to prove the rotation of the earth, under conditions as simple as at the poles; it is free from the objection raised against Foucault's gyroscope, viz. being connected with the earth by other points than its centre of gravity. We propose to show theoretically in a subsequent chapter, that by Foucault's mode of suspension the same effect is produced as by supporting the revolving body solely on its centre of gravity.

CHAPTER IX.

THE PRECESSION OF THE EQUINOXES A PROOF OF THE EARTH'S ROTATION.

THE ecliptic, i. e. the apparent path described by the sun during a year, cuts the equator in two opposite points, dividing it, and being divided by it in two equal parts. It has been observed from the earliest periods of antiquity, that when the sun was in either of these points of intersection, his diurnal revolution coincided with the equator, and therefore the days and nights were equal; hence the equator was called the equinoctial line, its intersections with the ecliptic the equinoctial points, one of them the vernal, the other the autumnal equinox.

It was evidently an important problem in practical astronomy to determine the exact moment of the sun's occupying these stations; for it was natural to compute the course of the year from that moment. Accordingly, this has been the leading problem in the astronomy of all nations. It is susceptible of considerable precision, without any apparatus of instruments. It is only necessary to observe the sun's declination on the noon of two or three days before and after the equinoctial day. On two consecutive days of this number, his declination must have changed from north to south, or from south to north. If his declination on one day was observed to be 21′ north, and on the next 5′ south, it follows that his declination was nothing, or that he was in the equinoctial point about twenty-three minutes after seven in the morning of the second day. Knowing the precise moment, and knowing the rate of the sun's motion in the ecliptic, it is easy to ascertain the precise point of the ecliptic in which the equator intersected it.

By a series of such observations made at Alexandria between the years 161 and 127 before Christ, Hipparchus, the great father of astronomy, found that the point of the autumnal equinox was about six degrees to the eastward of the star called *spica virginis.* Eager to determine everything by multiplied observations, he ransacked all the Chaldean, Egyptian, and other records, to which his travels could procure him access, for observations of the same kind; but he does not mention his having found any. He found, however, some observations of Aristillus and Timochares, made about 153 years before. From these it appeared evident that the point of the autumnal equinox was then about eight degrees east of the same star. He discusses these observations with great sagacity and vigour; and on their authority he asserts that the equinoctial points are not fixed in the heavens, but move to the westward about a degree in seventy-five years, or somewhat less.

This motion is called the *precession of the equinoxes,* because by it the time and place of the sun's equinoctial station precedes the usual calculations: it is fully confirmed by all subsequent observations. In 1750 the autumnal equinox was observed to be 20° 21′ westward of spica virginis. Supposing the motion to have been uniform during the previous period, it follows that the annual precession is about 50·3″; that is, if the celestial equator cuts the ecliptic in a particular point on any day of this year, it will on the same day of the following year cut it in a point 50·3″ to the west of it, and the sun will come to the equinox 20′ 23″ before he has completed his round of the heavens. Thus the equinoctial or tropical year, or true year of seasons, is so much shorter than the revolution of the sun or sidereal year.

It is this discovery that has chiefly immortalised the name of Hipparchus, though it must be acknowledged that all his astronomical researches have not been conducted with the same sagacity and intelligence. It was natural, therefore, for him to value himself highly on the discovery. It must be acknowledged to be one of the most singular that has been made, that the revolution of the whole heavens should not be stable, but its axis continually changing. For it must be observed that since the equator changes its position, and the equator is only an imaginary circle equidistant from the two poles or extremities of the axis, these poles and this axis must equally change their positions. The equinoctial points make a complete revolution in about 25,745 years, the equator being all the while inclined to the ecliptic in nearly the same angle. Therefore the poles of this diurnal revolution must describe a circle round the poles of the ecliptic at the distance of about 23½ degrees in 25,745 years; and in the time of Timochares the north pole of the heavens must have been thirty degrees eastward of where it now is.

Hipparchus has been accused of plagiarism and insincerity in this matter. It is now very certain that the precession of the equinoxes was known to the astronomers of India many ages before the time of Hipparchus. It appears also that the Chaldeans had a pretty accurate knowledge of the year of seasons. From their saros we deduce their measure of this year to be 365 days, 5 hours, 49 minutes, and 11 seconds, exceeding the truth only by twenty-six seconds, and much more exact than the year of Hipparchus. They had also a sidereal year of 365 days, 6 hours, 11 minutes. Now what could occasion an attention to two years, if they did not suppose the equinoxes movable? The Egyptians also had a knowledge of something equivalent to this; for they had discovered that the dog-star was no longer the faithful forewarner of the overflowing of the Nile, and they combined him with the Fomalhaset in their mystical calendar. This knowledge is also involved in the precepts of the Chinese astronomy, of much older date than the time of Hipparchus.

But all these acknowledged facts are not sufficient for depriving Hipparchus of the honour of the discovery, or fixing on him the charge of plagiarism. This motion was a thing unknown to the astronomers of the Alexandrian school, and it was pointed out to them by Hipparchus in the way in which he ascertained every other position in astronomy, namely, as the mathematical result of actual observations, and not as a thing deducible from any opinions on other subjects related to it. We see him on all other occasions eager to confirm his own observations, and his deductions from them, by everything he could pick up from other astronomers; and he even adduced the above-mentioned practice of the Egyptians in corroboration of his doctrine. It is more than probable then that he did not know anything more. Had he known the Indian precession of 54″ annually, he had no temptation whatever to withhold him from using it in preference to one which he acknowledges to be inaccurate, because deduced from the very short period of 150 years, and from the observations of Timochares in which he had no great confidence.

This motion of the heavens was long a matter of discussion as a phenomenon for which no physical reason could be assigned, but the Copernican system rendered it easy of explanation. A simple experiment with the ordinary spinning top will serve to show that a motion may exist about an axis which has itself a conical motion without, at the same time affecting its inclination. No more perfect illustration of astronomical precession can be derived than that presented by a properly balanced top which exhibits at the same time the other phenomena of rotation. In the experimental portion of our subject we have noticed Bohnenberger's rotatory apparatus, which is the model of the gyroscope by which Foucault made apparent the effects of the earth's rotation. We may here mention Troughton's nautical top, and Elliot's, which is similar to it, though an original invention; but we will confine ourselves to a description of Professor Maxwell's dynamical top, which differs from Elliot's in having more adjustments, and in being designed to exhibit far more complicated phenomena: there is also an essential addition in the coloured disc for showing the motion of the axis. The conditions which must be observed in the construction of such instruments are the following. The body must be balanced on its centre of gravity, and capable of having its principal axes and moments of inertia altered in form and position; the axle of the instrument must be capable of being made the greatest, least, or mean principal axis, and the position of the invariable axis of rotation must admit of being made visible at any time.

Fig. 29.

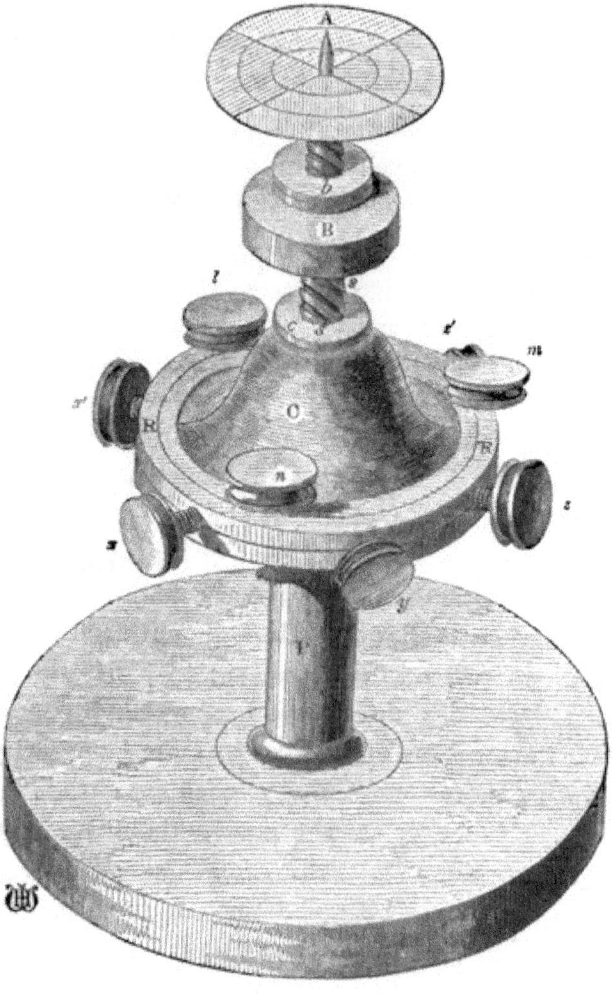

MAXWELL'S DYNAMICAL TOP.

There must be three adjustments to regulate the positions of the centre of gravity, three for the magnitudes of the moments of inertia, and three for the directions of the principal axes in all nicer independent adjustments which may be distributed *ad libitum* among the screws of the instrument.

The form of the body of the apparatus which Professor Maxwell has found most suitable is that of a bell, see fig. 29. C is a hollow cone of brass, R is a heavy ring cast in the same piece, six screws with heavy heads, $x\,y\,z$ and $x'\,y'\,z'$ work horizontally in the ring, and three similar screws, $l\,m\,n$, work vertically through the ring at equal intervals. A S is the axle of the instrument, $s\,s$ is a brass screw working the upper part of the cone C, and capable of being firmly clamped by means of the nut c. B is a cylindrical brass bob which may be screwed up or down the axle and fixed in the required place by the nut b. The lower extremity of the axle is a fine steel point finished without emery, and afterwards hardened; it runs in a little agate cup set in the top of the pillar P. The upper end of the axle has also a steel point by which it may be kept steady while spinning; a coloured disc is attached to it when the instrument is in use.

Dimensions and weights of the parts of the dynamical top.

I. Body of the top.

	lb.	oz.
Mean diameter of ring 4 inches.		
Section of ring ½ inch square.		
The conical portion rises from the upper and inner edge of the ring a height of 1½ inch		
The whole body of the top weighs . . .	1	7
Each screw is one inch long, and weighs one ounce, together		9

II. Axle, &c.	lb.	oz.
Length of axle 5 inches, of which ½ inch at the bottom occupied by the steel point, 3½ inches are brass with a screw turned on it, and the remaining inch is of steel with a sharp point at the top. The whole weighs		1½
The bob B has a diameter of 1·4″ and thickness of 0·4. It weighs		2¾
The nuts *b* and *c* for clamping the bob and the body of the top on the axle weigh . . .		1
Weight of whole top . . .	2	5¼

The disc of card is divided into seven quadrants, coloured with vermilion, chrome yellow, emerald green and ultra-marine. They combine into a greyish tint when the revolution is about the axle, and burst into brilliant colours when the axis is disturbed. It is useful to have some concentric circles, drawn with ink, over the colours, and about 12 radii drawn in strong pencil lines. It is easy to distinguish the ink from the pencil lines, as they cross the invariable axis by their want of lustre. In this way the path of the invariable axis may be identified with great accuracy and compared with theory.

The first thing to be done with the instrument is to make the steel point at the end of the axle coincide with the centre of gravity. The adjustment for it being made, the rotation of the top has no tendency to shake the steel point in the agate cup.

The next thing to be done, is to make one of the principal axes of the central ellipsoid coincide with the axle of the top. We effect this by spinning the top gently about its axle, steadying the upper part with the finger at first. If the axle is already a principal axis, the top will continue to revolve about its axle, when the finger is removed. If it is not, the top begins to spin about some other axis, alternately widening its circles and contracting them. The path which the invariable axis describes upon the disc may be a circle, an ellipse, a hyperbola, or a straight line according to the arrangement of the instrument.

In the case in which the invariable axis coincides at first with the axle of the top, and returns to it after separating from it, its true path is a circle or an ellipse, having the axle in its circumference. The true principal axis is at the centre of the closest curve. It must be made to coincide with the axle, by adjusting the vertical screws l, m, n.

Let us suppose that the principal axis is that of the greatest moment of inertia, and that we have made it coincide with the axle of the instrument. Let us also suppose that the moments of inertia about the other axes are equal, and very little less than that about the axle. Let the top be spun about the axle, and then receive a disturbance which causes it to spin about some other axis. The instantaneous axis will not remain at rest, either in space or in the body. In space it will describe a right cone, completing a revolution, in somewhat less than the time of revolution of the top. In the body it will describe another cone of larger angle, in a period which increases inversely with the difference of axes of the body. The invariable axis will be fixed in space, and describe a cone in the body.

Another small and slow movement of the earth's axis, which is dependent on the variable position of the lunar orbit, afford the most cogent proofs that our globe is endowed with rotation, otherwise we could only account for the phenomena it produces by the absurd hypothesis that the whole celestial sphere is kept in a constant state of tremor by the influence of the moon. The small but important inequality we speak of, is called "Nutation," a name given to it by its discoverer, Dr. Bradley, in a memoir communicated to the Royal Society in 1748. It is a kind of tremulous or nodding motion of the earth's axis, whereby its inclination to the plane of the ecliptic varies backwards and forwards to the amount of some seconds. By this motion, if subsisting alone, the pole would describe among the stars, in the period of about nineteen years, a small ellipse, with axes respectively 18″ 5 and 13″ 75, but in conjunction with the motion of precession, the path pursued will be a gently undulated ring.

The two motions of precession and nutation are consequences of the earth's rotation, and ellipticity, combined with the unequal attraction of the sun and moon on its polar and equatorial regions. In order to exhibit these motions jointly, and to illustrate the causes which produce them, an ingenious instrument was invented by Mr. Burr, of which the following is a description. See fig. 30.

It consists of a small globe representing the earth as seen from the north pole, from which a small axle of steel projects, while the equator is widened into a broad metallic ring; the globe is supported at a point within it, to which an axle descends from the foot of the apparatus, which at the same time carries an horizontal ring, representing the ecliptic. The whole is so balanced, that in a state of rest, the equator and ecliptic coincide. If a small weight be attached to the ring of the equator, representing the attraction of the sun and moon, the axis

FIG. 30.

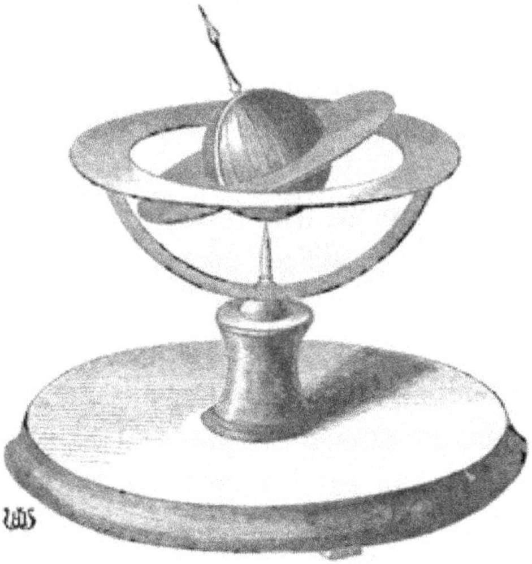

BURR'S APPARATUS FOR THE IMITATION OF PRECESSION.

will be inclined, and the latter being put in quick rotation, it will be observed that the projecting axle is describing a small motion round the pole of the ecliptic in the surface of a cone, and in a direction contrary to that of the earth's rotation, in consequence of which, a shifting of the equinoctial points (at which the equator and ecliptic intersect), is produced in a retrograde direction. The phenomenon of the precession of the equinoxes is hereby clearly imitated; and more, if the rotation of the projecting axle be closely and attentively watched, it will be observed to undergo a number of oscillations in its path round the pole of the ecliptic — oscillations which exactly correspond with the nutation of the earth.

These motions are easily explained by the first principles of mechanics. A small weight on the plane of the equator distributes the mass of the apparatus unequally with regard to the axis of rotation, in consequence of which centrifugal forces are produced, the resultant of which does not pass through the point of suspension, viz. the centre of the globe.

On the other hand, the globe being hollowed out at the bottom in order to fit in the lower axle, it has more mass in the upper than in the lower part, the centre of gravity and the geometrical centre of the globe do not coincide, and thence again a resultant of centrifugal forces not passing through the centre is the natural consequence.

The least inclination of the axle, without any overweight at the equator, is sufficient to produce the phenomenon of precession, while the small weight attached to it acts as a second disturbing force, thus producing the phenomenon of nutation as observed in the minute oscillations of the axle or nodding motion in its circular path round the pole of the ecliptic.

We cannot do better than conclude this, the first part of our work, with the words of Sir John Herschel: "These movements of 'precession' and 'nutation' are common to all the celestial bodies both fixed and erratic, and this circumstance makes it impossible to attribute them to any other cause than a real motion of the earth's axis such as we have described."

PART II.

PART II.

CHAPTER I.

THEORY OF THE MOTION OF FALLING BODIES, BY LA PLACE AND GAUSS.

IN the following analysis, La Place gives the true expression for the deviation of falling bodies from the vertical, taking into consideration the resistance of the air, and showing that whatever that resistance may be, and whatever may be the form of the earth, no deviation takes place towards the equator.

Let $x\ y\ z$ be the three rectangular co-ordinates of the body, the origin of co-ordinates being the centre of the earth, and the axis of x its axis of rotation. Let r be the radius drawn from the centre to the summit of the tower from which the body falls; θ the angle which r makes with the axis of rotation, and a the angle which the plane passing through r, and through the axis of the earth, makes with the plane passing through the same axis, and also through one of the principal axes of the earth, situated in the plane of its equator; and let $n\ t$ be the angular motion of rotation of the earth. Taking X Y Z, the co-ordinates of the summit of the tower, we have

$$X = r \cos \theta$$
$$Y = r \sin \theta \cos (n\,t + \omega)$$
$$Z = r \sin \theta \cos (n\,t + \omega)$$

$n\,t + \omega$ being the angle which the plane passing through r and through the axis of the earth makes with the plane of x and y. Let us now suppose that, relatively to the falling body, r changes itself into $s - a\,s$, θ into $\theta + a\,u$, and ω into $\omega + a\,v$, then,

$$x = (r - as)\cos(\theta + au)$$
$$y = (r - as)\sin(\theta + au)\cos(n\,t + \omega + a\,v)$$
$$z = (r - as)\sin(\theta + au)\sin(n\,t + \omega + a\,v)$$

Let V be the sum of all the particles composing the terrestrial spheroid, divided by their distances from the body attracted; the forces which act on this body, by the attraction of these particles, are parallel to the axes of $x\ y\ z$

$$\left(\frac{d\,\mathrm{V}}{dx}\right)\ \left(\frac{d\,\mathrm{V}}{dy}\right) \text{ and } \frac{d\,\mathrm{V}}{dz}$$

(see No. 21, Vol. II. of the "Mécanique Céleste"). In order to consider the resistance of the air, we may represent it by the expression

$$\phi\left(a\ s\ a\,\frac{d\,s}{d\,t}\right)$$

For, the speed of the body relatively to the air, considered at rest, is considerably greater in the direction of r than in that perpendicular to r; we shall see that the expression for this relative speed is approximately

$$a\,\frac{d\,s}{d\,t}$$

If now, for simplicity, we make

$$r = 1$$

then the relative speed of the body in the direction of θ is $a\,\frac{d\,u}{d\,t}$ and in the direction ω it is $a\,\frac{d\,v}{d\,t}\sin\theta$.

The resistance of the air is therefore $\dfrac{\left(a s\ a\,\frac{d\,s}{d\,t}\right)}{a\,\frac{d\,s}{d\,t}}\ a\,\frac{d\,s}{d\,t}$ in the direction of r,

$$\frac{-\phi\left(a s\,\alpha\,\frac{ds}{dt}\right)}{\alpha\,\frac{ds}{dt}}\,\alpha\,\frac{du}{dt}\ \text{in the direction of}\ \theta$$

$$\frac{-\phi\left(a s\,\alpha\,\frac{ds}{dt}\right)}{\alpha\,\frac{ds}{dt}}\,\alpha\,\frac{dv}{dt}\sin\theta\ \text{in the direction of}\ \omega.$$

Let K represent the factor $\frac{\phi\left(a s\,\alpha\,\frac{ds}{dt}\right)}{\alpha\,\frac{ds}{dt}}$, then we have, by the principle of virtual velocities,

$$0=\delta x\,\frac{ddx}{dt^2}+\delta y\,\frac{ddy}{dt^2}+\delta z\,\frac{ddz}{dt^2}-\delta x\left(\frac{dV}{dx}\right)-\delta y\left(\frac{dV}{dy}\right)$$
$$-\delta z\left(\frac{dV}{dz}\right)-K\,\delta r\,\alpha\,\frac{ds}{dt}+K\,\delta\theta\,\alpha\,\frac{du}{dt}+K\,\delta\omega\,\sin^2\theta\,\alpha\,\frac{dv}{dt}$$

the differential δ relating to the co-ordinates r, θ, and ω, of which $x\ y\ z$ are functions. Substituting for $x\ y\ z$ their preceding values, and neglecting the terms of α^2, we have

$$0=\delta r\left(-\alpha\,\frac{dds}{dt^2}-2\,\alpha\,n\,r\,\frac{dv}{dt}\,\sin^2\theta-\alpha\,K\,\frac{ds}{dt}\right)$$
$$+r^2\,\delta\,\theta\left(\alpha\,\frac{ddu}{dt^2}-2\,\alpha\,n\,\frac{dv}{dt}\,\sin\theta\,\cos\alpha+\alpha K\,\frac{du}{dt}\right)$$
$$r^2\,\delta\,\omega\,\sin\theta\left(\alpha\,\frac{ddv}{dt^2}\,\sin\theta+2\,\alpha\,n\,\frac{du}{dt}\,\cos\theta-2\,\alpha\,n\,\frac{ds}{dt}\,\frac{\sin\theta}{r}+\alpha\,K\,\frac{dv}{dt}\sin\theta\right)$$
$$-\delta\,V-\frac{n^2}{2}\,\delta\left((r-\alpha s)^2\,\sin^2(\theta+\alpha u)\right)\qquad(1)$$

By the nature of the equilibrium of the stratum of air in which the body is, we have

$$0=\delta\,V+\frac{n^2}{2}\,\delta\left((r-\alpha s)^2\,\sin^2(\theta+\alpha u)\right)\qquad(2)$$

(vide "Mécanique Celeste," Vol. I. p. 98),

provided that the value of $\delta\,r$ be considered in relation to the level surface of the stratum of air. Take on this surface $r=a+y$; y being a function of θ, of ω, and of a, a being constant for the same stratum, then by equation (2), we obtain

$$0=\frac{dQ}{dr}\left(\left(\frac{dy}{d\theta}\right)\delta\theta+\left(\frac{dy}{d\omega}\right)\delta\omega\right)+\left(\frac{dQ}{d\theta}\right)\delta\theta+\left(\frac{dQ}{d\omega}\right)\delta\omega$$

Q being supposed $=$ to $V+\frac{n}{2}\left(\left(r-\alpha s\right)\sin^2\left(\theta+\alpha u\right)\right)$

Subtracting this equation from equation (1) we have

$$0=\delta r\left(-a\,\frac{dds}{dt^2}\,2\,anr\frac{dv}{dt}\sin^2\theta-a\,K\,\frac{ds}{dt}\right)$$

$$+r^2\delta\theta\left(a\,\frac{ddu}{dt^2}\,-2\,an\frac{dv}{dt}\sin\theta\cos\theta+a_K\,\frac{du}{dt}\right)$$

$$+r^2\,\delta\,\omega\,\sin\theta\left(a\,\frac{ddv}{dt^2}\sin\theta+2\,an\frac{du}{dt}\,\cos\theta-2\,an\,\frac{ds}{dt}\frac{\sin\theta}{r}\,+a\,K\,\frac{dv}{dt}\sin\theta\right)$$

$$-\frac{dQ}{dt}\left(\delta r-\left(\frac{dy}{d\theta}\right)\delta\theta-\left(\frac{dy}{d\omega}\right)\delta\omega\right).$$

If now we take the coefficients of the three variations δr, $\delta\theta$, and $\delta\omega$ respectively equal to 0, and observe that $\frac{dQ}{dt}$ represents the gravity which we designate g (vide "La Mécanique Céleste," Vol. II. p. 104), we shall obtain, taking the radius r as unity (which may be done without appreciable error), the three following equations :—

$$0=a\frac{dds}{dt^2}+2\,an\,\frac{dv}{dt}\,\sin^2\theta+a\,K\,\frac{ds}{dt}-g$$

$$0=a\frac{ddu}{dt^2}-2\,an\frac{dv}{dt}\,\sin\theta\cos\theta+a\,K\,\frac{du}{dt}-g\left(\frac{dr}{d\theta}\right)$$

$$0=a\frac{ddv}{dt^2}\,\sin\theta+2\,an\,\frac{du}{dt}\,\cos\theta-2\,an\,\frac{ds}{dt}\sin\theta$$

$$+\,a\,K\,\frac{dv}{dt}\sin\theta-\,\frac{g}{\sin\theta}\left(\frac{dy}{d\omega}\right)$$

If we take the second decimal, or the $\frac{1}{100000}$ of the mean day as the unity of time, n is the small angle described in one second by the rotation of the earth. This angle is exceedingly small, and as $a\,u$ and $a\,v$ are very minute quantities in relation to $a\,s$, we may neglect in the first of these three equations the term

$$2\,a\,n\,\frac{dv}{dt}\sin^2\theta$$

in the second

$$2\,a\,n\,\frac{dv}{dt}\sin\theta\cos\theta$$

and in the third

$$2\,a\,n\,\frac{du}{dt}\cos\theta$$

which reduces them to the following:—

$$0 = a\frac{dds}{dt^2} + a\,\mathrm{K}\frac{ds}{dt} - g$$

$$0 = a\frac{ddu}{dt^2} + a\,\mathrm{K}\frac{du}{dt} - g\left(\frac{dy}{d\theta}\right)$$

$$0 = a\frac{ddv}{dt^2}\sin\theta - 2\,a\,n\frac{ds}{dt}\sin\theta + a\,\mathrm{K}\frac{dv}{dt}\sin\theta - \frac{g}{\sin\theta}\left(\frac{dy}{d\omega}\right)$$

K being the function of $a\,s$ and of $a\,\frac{ds}{dt}$, the first of these equations gives $a\,s$ in functions of the time t. The second of these equations can be satisfied by making

$$a\,u = a\,s\left(\frac{dy}{d\theta}\right)$$

because g and $\frac{dy}{d\theta}$ may be supposed constant through the duration of the movement, from the smallness of the height from which the body falls, as compared with the radius of the earth. This is the only manner of satisfying the second equation in the present analysis, for u, $\frac{du}{dt}$, s, and $\frac{ds}{dt}$ are nought at the commencement of the movement.

If now we imagine a plumb-line of the length $a\,s$ suspended from the point whence the body falls, it will deviate towards the south from the radius r, by the quantity

$$a\,s\left(\frac{dv}{d\theta}\right)$$

and consequently by the quantity $a\,u$; the body in falling is always on the parallels of the points of the vertical, which are at the same height as itself; it does not diverge to the south of this line.

In order to integrate the third equation we will make

$$a v \sin\theta = \frac{a s}{\sin\theta}\frac{d y}{d\omega} + a v'$$

and we have

$$0 = a\frac{d d v'}{d t^2} + a\,\mathrm{K}\frac{d v'}{d t} - 2 a n \frac{d s}{d t}\sin\theta$$

The body deviates to the east from the radius r by the quantity:

$$a v \sin\theta = \frac{a s}{\sin\theta}\left(\frac{d y}{d\omega}\right) + a v'$$

but the plumb line deviates to east from the same radius by the quantity

$$\frac{a s}{\sin\theta}\left(\frac{d y}{d}\right)$$

$a\,v'$ is therefore the deviation of the body to the east of the vertical.

Let us now suppose the resistance of the air to be proportional to the square of the velocity, so that

$$\mathrm{K} = m a \frac{ds}{dt}$$

m being a coefficient depending on the form of the body, but which we may here, without appreciable error, consider constant.

Then we have

$$0 = a\frac{dds}{dt^2} + a^2 m \frac{ds^2}{dt^2} - g$$

In order to integrate this equation we will make

$$as = \frac{1}{m}\log.\, s'$$

then

$$0=\frac{dds}{dt^2}-mgs'$$

which by integration gives

$$s'=\mathrm{A}c^{t\sqrt{mg}}+\mathrm{B}c^{-t\sqrt{mg}}$$

c being the number the hyperbolic logarithm of which is unity, and A and B being two arbitrary quantities. In order to determine them, we must remark that αs should be nought when $t=0$, which makes $s'=1$; therefore,

$$\mathrm{A}+\mathrm{B}=1.$$

Besides $\alpha\frac{ds}{dt}$ must be nought with t, consequently with $\frac{ds'}{dt}$ which gives

$$\mathrm{A}-\mathrm{B}=0$$

therefore $\mathrm{A}=\mathrm{B}=\frac{1}{2}$, consequently

$$-\alpha s=\frac{1}{m}\log.\left(\tfrac{1}{2}c^{t\sqrt{mg}}+\tfrac{1}{2}c^{-t\sqrt{mg}}\right)$$

and reducing into a series,

$$\alpha s=\frac{g\,t^2}{2}-\frac{m\,g^2t^4}{12}+\frac{m^2g^3t^6}{45}-\ \&\text{c}.$$

In order to determine $\alpha v'$, we must consider that

$$\alpha\frac{ds}{dt}=\frac{1}{m}\frac{ds'}{s'dt}$$

and that therefore the differential equation in $\alpha v'$ becomes

$$0=\alpha s'\frac{ddv'}{dt^2}+\alpha\frac{ds'}{dt}\frac{dv'}{dt}-\frac{2n}{m}\frac{ds'}{dt}$$

from which we obtain by integration,

$$\alpha s'\frac{dv'}{dt}=\frac{2n}{m}s'+\mathrm{C}$$

C being an arbitrary constant, in order to determine which,

we must remember that t being nought $\frac{d v'}{d t} = 0$, and that then $s' = 1$ which makes

$$C = -\frac{2n}{m}$$

$\therefore$ dividing by s' and substituting the value of C

$$a \frac{dv'}{dt} = \frac{2n}{m}\left(1 - \frac{1}{s'}\right) = \frac{2n}{m}\left(1 - \frac{a}{c^{t\sqrt{mg}} + c^{-t\sqrt{mg}}}\right)$$

integrating so that $a\,v'$ be nought with t we have

$$av' = \frac{2n}{m}t - \frac{4n}{m\sqrt{mg}} \operatorname{tang}^{-1}\left(\frac{c^{\frac{t}{2}\sqrt{mg}} - c^{-\frac{t}{2}\sqrt{mg}}}{c^{\frac{t}{2}} + c^{-\frac{t}{2}\sqrt{mg}}}\right)$$

and reducing into a series

$$av' = \frac{ngt^3 \sin\theta}{3}\left(1 - \frac{mgt^2}{4} + \frac{61}{840} m^2g^2t^4 - \&c.\right)$$

We must consider in these expressions of $a\,s$ and $a\,v'$, that t, expressing a number of units of time, g is double the space caused by gravity in the first unit of time, $n\,t$ is the angle of rotation of the earth during t unities, and $m\,g$ is a number depending on the resistance offered by the air to the movement of the body. In order to obtain the time of falling and the deviation towards the east in functions of the height from which the body falls, let us represent the height by h; then, by the preceding, we have

$$2c^{mh} = c^{t\sqrt{mg}} + c^{-t\sqrt{mg}}$$

whence

$$t = -\frac{1}{\sqrt{mg}} \quad \frac{1}{2}\left(\sqrt{c^{mh}+1} + \sqrt{c^{mh}-1}\right)$$

and

$$av' = \frac{2n}{m\sqrt{mg}}\left(\log \frac{1}{2}\left(\sqrt{c^{\sqrt{mh}}+1} + \sqrt{c^{\sqrt{mh}}-1}\right)^2 - 2\operatorname{tang}^{-1}\left(\frac{\sqrt{c^{\sqrt{mh}}-1}}{\sqrt{c^{\sqrt{mh}}+1}}\right)\right)$$

The height h being given, the observation of the time t

will give the value of m, and we shall derive $a v'$, or the deviation of the body towards the east of the vertical. The coincidence of this theoretical result with the practical experiments, renders the rotary motion of the earth manifest; m could also be determined by the figure and density of the body, and by experiments already made on the resistance of the air.

In vacuo, or which is the same thing, when m is excessively small, we have

$$a v' = \frac{2 n h}{3} \sin \theta \frac{\sqrt{2 h}}{g}$$

θ is very nearly the complement of the latitude of the place, and for Paris we may suppose $\theta = 41° 9' 46''$, n is the angle of rotation of the earth during a unit of time. If now we take for this unit the $\frac{1}{100000}$ part of a day, we shall obtain

$$n = \frac{1296000}{99727}$$

because the duration of the earth's rotation is 0·99727 of a day, and for Paris,

$$\tfrac{1}{2} g = 3{,}66107 \text{ mètres.}$$

Supposing, therefore,

$$h = 54 \text{ mètres,}$$

we have

$$a v' = 5{,}7337 \text{ millimètres.}$$

The following are the fundamental formulæ for the motion of a falling body with regard to the rotation of the earth, as developed by Gauss. The position of a point may be determined in two ways:—

1st. By its three perpendicular distances, X, Y, Z, from three fixed planes perpendicular to each other. Their common point of intersection C is placed in the earth's axis, the plane of Z is parallel to the equator, the plane of

Y in that meridian which passes through the original position of the body, and the plane of X in the meridian perpendicular to the former. The co-ordinates of Z are positive towards the north, those of X towards the primitive place of the body, and those of Y towards that side to which the first place is carried by rotation.

2ndly. By the three perpendicular distances x, y, z, from three movable planes, which relatively to the earth are at rest, but rotate with it; their common point of intersection is best taken in the primitive place of the body.

The plane of z is perpendicular on the direction of gravity, the plane of y lies in the meridian, and the third plane is perpendicular to the two former. The poles of these three planes are respectively the zenith, the east point, and the south point, towards which the co-ordinates z, y, x, are taken positive.

Let for the point C, $x=a$, $y=0$, $z=-c$, and the complement of the latitude of the place of observation $=\phi$, and let the angle of the earth's rotation during the time t be denoted by θ, then we have the following equations:—

$$\left.\begin{array}{l} x=\mathrm{X}\sin\phi\cos\theta+\mathrm{Y}\sin\phi\sin\theta-\mathrm{Z}\cos\phi+a \\ y=\mathrm{Y}\sin\theta+\mathrm{Y}\cos\theta \\ z=\mathrm{X}\cos\phi\cos\theta+\mathrm{Y}\cos\phi\sin\theta+\mathrm{Z}\sin\phi-c \end{array}\right\} \quad . \quad (1)$$

$$\left.\begin{array}{l} \mathrm{X}=(x-a)\sin\phi\cos\theta-y\sin\theta+(z+c)\cos\phi\cos\theta \\ \mathrm{Y}=(x-a)\sin\phi\sin\theta+y\cos\theta+(z+c)\cos\phi\sin\theta \\ \mathrm{Z}=-(x-a)\cos\phi+(z+c)\sin\phi \end{array}\right\} \quad . \quad (2)$$

The co-ordinates X, Y, Z, may be regarded either as functions of t, or as those of the four variable quantities θ, x, y, z, and with regard to the latter they have four partial differentials. Therefore

$$d\mathrm{X}=\left(\frac{d\mathrm{X}}{dt}\right)dt=\left(\frac{d\mathrm{X}}{d\theta}\right)d\theta+\left(\frac{d\mathrm{X}}{dx}\right)dx+\left(\frac{d\mathrm{Y}}{dy}\right)dy+\left(\frac{d\mathrm{Z}}{dz}\right)dz$$

The velocity of the body may be resolved into three, the directions of which are perpendicular upon the planes of X, Y, Z, and are therefore expressed by

$$\left(\frac{dX}{dt}\right), \left(\frac{dY}{dt}\right), \left(\frac{dz}{dt}\right)$$

The velocities of the particle of air in which the centre of gravity of the body is at any moment, referred to the same axes, are

$$\left(\frac{dX}{d\theta}\right) d\theta, \left(\frac{dY}{d\theta}\right) d\theta, \left(\frac{dZ}{d\theta}\right) d\theta$$

hence the relative velocities of the body in these three directions,

$$\xi = \left(\frac{dX}{dx}\right)\frac{dx}{dt} + \left(\frac{dX}{dy}\right)\frac{dy}{dt} + \left(\frac{dX}{dz}\right)\frac{dz}{dt}$$

$$= \sin\phi \cos\theta \frac{dx}{dt} - \sin\theta \frac{dy}{dt} - \cos\phi \cos\theta \frac{dz}{dt}$$

$$\eta = \left(\frac{dY}{dx}\right)\frac{dx}{dt} + \left(\frac{dY}{dy}\right)\frac{dy}{dt} + \left(\frac{dY}{dz}\right)\frac{dz}{dt}$$

$$= \sin\phi \sin\theta \frac{dx}{dt} + \cos\theta \frac{dy}{dt} + \cos\phi \sin\theta \frac{dz}{dt}$$

$$\zeta = \left(\frac{dZ}{dx}\right)\frac{dx}{dt} + \left(\frac{dZ}{dy}\right)\frac{dy}{dt} + \left(\frac{dZ}{dz}\right)\frac{dz}{dt}$$

$$= - \cos\phi \frac{dx}{dt} + \sin\phi \frac{dz}{dt}$$

The total relative velocity is therefore

$$u = \sqrt{\xi^2 + \eta^2 + \zeta^2} = \sqrt{\frac{dx^2}{dt^2} + \frac{dy^2}{dt^2} + \frac{dz^2}{dt^2}}$$

The square of this quantity is proportional to the resistance of the air, which is therefore expressed by Mu, and its three components in the directions above stated are

$$Mu\xi, \; Mu\eta, \; Mu\zeta,$$

Considering the earth as a spheroid of revolution, the direction of gravity will pass through the earth's axis. Let the point where these two lines intersect be above C by the quantity q, i. e. its co-ordinate $Z=q$, taking the force of gravity$=p$, and $X^2=Y^2+(Z-q)^2=r^2$. We have, by the principles of Dynamics,

$$\left.\begin{aligned} 0&=\frac{d^2X}{dt^2}+\frac{pX}{r}+Mu\xi \\ 0&=\frac{d^2Y}{dt^2}+\frac{pY}{r}+Mu\eta \\ 0&=\frac{d^2Z}{dt^2}+\frac{p(Z-q)}{r}+Mu\zeta \end{aligned}\right\} \quad . \quad . \quad . \quad (3)$$

From the values X, Y, Z, in (2) putting for the constant $\frac{d\theta}{dt}=n$, we obtain the following equations:—

$$\left.\begin{aligned} \frac{d^2X}{dt^2} &= \sin\phi\cos\theta\frac{d^2x}{dt^2}-\sin\theta\frac{d^2y}{dt^2}+\cos\phi\cos\theta\frac{d^2z}{dt^2} \\ &-2n\sin\phi\sin\theta\frac{dx}{dt}-2n\cos\theta\frac{dy}{dt}-2n\cos\phi\sin\theta\frac{dz}{dt}-n^2X \\ \frac{d^2Y}{dt^2} &= \sin\phi\sin\theta\frac{d^2x}{dt^2}+\cos\theta\frac{d^2y}{dt^2}+\cos\phi\sin\theta\frac{d^2z}{dt^2} \\ &+2n\sin\phi\cos\theta\frac{dx}{dt}-2n\sin\theta\frac{dy}{dt}+2n\cos\phi\cos\theta\frac{dz}{dt}-n^2Y \\ \frac{d^2Z}{dt^2} &= -\cos\phi\frac{dx^2}{dt^2}+\sin\phi\frac{d^2z}{dt^2} \end{aligned}\right\} (4)$$

Multiplying the three equations (3) respectively by $\sin\phi\cos\theta$, $\sin\phi\sin\theta$, $-\cos\phi$, and adding the products; multiplying again these equations by $-\sin\theta$, $\cos\theta$, 0, and thirdly by $\cos\phi\cos\theta$, $\cos\phi\sin\theta$, $\sin\phi$, and adding each time the products, we obtain after substituting the values of ξ, η, ζ, as also those of $\frac{d^2X}{dt^2}$, $\frac{d^2Y}{dt^2}$, $\frac{d^2Z}{dt^2}$ from (4) and those X, Y, Z, from (2) the following three equations:—

$$\frac{d^2x}{dt^2} - 2n\sin\phi\frac{du}{dt} + \left(x-a\right)\left(\frac{p}{r} - n^2\right) + \cos\phi\left(\frac{pq}{r} - n^2 Z\right)$$
$$+ Mu\frac{dx}{dt} = 0$$

$$\frac{d^2y}{dt^2} + 2n\sin\phi\frac{dx}{dt} + 2n\cos\phi\frac{dz}{dt} + y\left(\frac{p}{r} - n^2\right) Mu\frac{dy}{dt} = 0$$

$$\frac{d^2z}{dt^2} - 2n\cos\phi\frac{dy}{dt} + (z+c)\left(\frac{p}{r} - n^2\right) + \sin\phi\left(\frac{nq}{r} - n^2 z\right)$$
$$+ Mu\frac{d^2}{dz} = 0.$$

If, therefore, the body is at rest with regard to the earth, or if $dx = dy = dz = 0$, then it is acted upon perpendicularly to the planes of x, y, z, by the following forces:—

$$\left.\begin{array}{l} (x-a)\left(\frac{p}{r} - n^2\right) + \cos\phi\left(\frac{pq}{r} - n^2 Z\right) \\ y\left(\frac{P}{r} - n^2\right) \\ (z+c)\left(\frac{p}{r} - n^2\right) + \sin\phi\left(\frac{nq}{r} - n^2 Z)\right) \end{array}\right\} \quad . \quad . \quad . \quad (5)$$

But a body already in motion is affected another way; for besides the resistance of the air, which urges it in three directions, like forces measured by $Mu\frac{dx}{dt}$, $Mu\frac{dy}{dt}$, $Mu\frac{dz}{dt}$, there are three other forces acting in the same directions, viz. $-2n\sin\phi\frac{dy}{dt}$, $2n\sin\phi\frac{dx}{dt} + 2n\cos\phi\frac{dz}{dt}$, $-2n\cos\phi\frac{dy}{dt}$, and these serve in falling bodies to show the rotation of the earth. As in all experiments of this kind, the space described by the body is comparatively very small, we may be allowed to assume the force of gravity acting upon it as a constant $= g$, and its directions always parallel, i. e.

perpendicular to the plane of z. On this assumption, we may substitute for the three quantities (5) respectively 0, 0, g; by this means the fundamental formulæ are transformed into the following :—

$$\frac{d^2x}{dt^2}-2n\sin\phi\frac{dy}{dt}+\mathrm{M}u\frac{dx}{dt}=0$$

$$\frac{d^2y}{dt^2}+2n\sin\phi\frac{dx}{dt}+2n\cos\phi\frac{dz}{dt}+\mathrm{M}u\frac{dy}{dt}=0$$

$$\frac{d^2z}{dt^2}-2n\cos\phi\frac{dy}{dt}+g+\mathrm{M}u\frac{dz}{dt}=0.$$

Neglecting the resistance of the air, i. e. taking M=0, we obtain by integration,—

$$\mathrm{X}=\mathrm{A}-\mathrm{D}\cos\phi, t+\mathrm{E}\frac{\sin\phi}{2n}\cos(2nt+\mathrm{F})+\tfrac{1}{2}\sin\phi\cos\phi\, gt^2$$

$$y=\mathrm{B}-\frac{\mathrm{E}}{2n}\sin(2nt+\mathrm{F})+\frac{\cos\phi\, gt}{2n}$$

$$z=\mathrm{C}+\mathrm{D}\sin\phi, t+\mathrm{E}\frac{\cos\phi}{2n}\cos(2nt+\mathrm{F})-\tfrac{1}{2}\sin\phi\, gt^2.$$

Granting that the initial velocity of the falling body was nought, then the values of the arbitrary quantities are easily determined, viz.—

$$\mathrm{A}=-\frac{g\cos\phi\sin\phi}{4n^2},\quad \mathrm{B}=0,\quad \mathrm{C}=-\frac{g\cos^2\phi}{4n^2}$$

$$\mathrm{D}=0,\ \mathrm{E}=\frac{g\cos\phi}{2n},\ \mathrm{F}=0$$

∴ by substitution,—

$$x=\frac{g\cos\phi\sin\phi}{2n}\left(nt^2-\frac{1}{2n}+\frac{\cos 2nt}{2n}\right)$$

$$y=\frac{g\cos\phi}{2n}\left(t-\frac{\sin 2nt}{2n}\right)$$

$$z=-\tfrac{1}{2}gt^2+\frac{g\cos^2\phi}{2n}\left(nt^2-\frac{1}{2n}+\frac{\cos 2nt}{2n}\right)$$

This integration is only true so long as the body does not move far from its original place; but if the distance be appreciable, we resolve the circular functions into series, and thus we obtain—

$$\left.\begin{array}{l} x=\frac{1}{6}\cos\phi\sin\phi\, g n^2 t^4 \\ y=\frac{1}{3}\cos\phi\, g n t^3 \\ z=\frac{1}{2} g t^2+\frac{1}{6}\cos^2\phi\, g n^2 t^4 \end{array}\right\} \quad (6)$$

As the time of falling lasts only a few seconds, the quantity $n t$ cannot exceed a few minutes of space, therefore x and the second term of small z become inappreciable; whence

$$y=-\tfrac{2}{3} z \cos\phi\, n t$$

Taking into consideration the resistance of the air, we must be satisfied with the approximate value of the integrals; and developing the expressions of x, y, z, in series according to the ascending powers of n, M, we get for the highest term of x, as before, $\frac{1}{6}\cos\phi\, g n^2 t^4$, which may be neglected, and for y and z we get, omitting the second and higher powers of n and M, the following values:—

$$\begin{aligned} y&=\tfrac{1}{3}\cos\phi\, g n t^3-\tfrac{1}{12}\cos\phi\, M g^2 n t^5 \\ z&=-\tfrac{1}{2} g t^2+\tfrac{1}{12} M g^2 t^4 \end{aligned}$$

Assuming the actual descent $-z=f$, and that in vacuo $\frac{1}{2} g t^2=f+\delta$, we have

$$\begin{aligned} y&=\tfrac{2}{3}\cos\phi\, n t(f+\delta)-\cos\phi\, n t\delta \\ &=\tfrac{2}{3}\cos\phi\, n t\left(f-\tfrac{1}{2}\delta\right) \end{aligned}$$

a formula which we obtained in Chapter V., Part I., by a more elementary method.

CHAP. II.

THEORY OF THE MOTION OF THE FREE PENDULUM WITH REGARD TO THE EARTH'S MOTION.

THE simple pendulum. Let x, y, z be three rectangular axes, the former two parallel to the equator of the earth, the third to its axis of rotation. Denoting by x, y, z the co-ordinates of the heavy particle, by x', y', z' those of the point of suspension, and by l the constant length of the pendulum, we have

$$(x-x')^2+(y-y')^2+(z-z')^2=l^2.$$

We may consider the material point as in free motion, if we introduce some indeterminate force μ acting upon it in a direction towards the point of suspension. The components of this force are, —

$$-\frac{\mu}{l}(x-x')\,;\,-\frac{\mu}{l}(y-y');-\frac{\mu}{l}(z-z') \quad . \quad . \quad (1)$$

Let V express the attraction of the earth, then the components of gravity are, —

$$\left(\frac{d\,\mathrm{V}}{d\,x}\right);\ \left(\frac{d\,\mathrm{V}}{d\,y}\right);\ \left(\frac{d\,\mathrm{V}}{d\,z}\right)$$

and if n denote the angular velocity of rotation, and x'', y'', z'' the co-ordinates of the centre of the earth, we have for the components of the centrifugal force, —

$$n^2(x-x'')\,;\ n^2(y-y'')\,;\ 0$$

hence, by subtraction, the components of attractive force acting upon the pendulum are, —

$$\left(\frac{d\,V}{d\,x}\right) - n^2(x-x''); \quad \left(\frac{d\,V}{d\,y}\right) - n^2(y-y''); \quad \left(\frac{d\,V}{d\,z}\right) . \quad (2)$$

The expressions (1) and (2) for the components of the accelerating forces acting upon the pendulum give, by addition, —

$$\frac{d^2x}{d\,t^2} = \left(\frac{d\,V}{d\,x}\right) - n^2(x-x'') - \frac{\mu}{l}(x-x')$$
$$\frac{d^2y}{d\,t^2} = \left(\frac{d\,V}{d\,y}\right) - n^2(y-y'') - \frac{\mu}{l}(y-y')$$
$$\frac{d^2z}{d\,t^2} = \frac{d\,V}{d\,z} \qquad\qquad - \frac{\mu}{l}(z-z')$$

Let x, y, z be the co-ordinates of the pendulous mass referred to the point of suspension, so that,

$$(x-x') = X; \quad (y-y') = y; \quad (z-z') = Z$$

then we obtain by substitution,—

$$\left.\begin{aligned} \frac{d^2X}{d\,t^2} &= \left(\frac{d\,V}{d\,x}\right) - n^2X - n^2(x'-x'') - \frac{d^2x'}{d\,t^2} - \frac{\mu}{l}\,X \\ \frac{d^2Y}{d\,t^2} &= \left(\frac{d\,V}{d\,y}\right) - n^2Y - n^2(y'-y'') - \frac{d^2y'}{d\,t^2} - \frac{\mu}{l}\,Y \\ \frac{d^2Z}{d\,t^2} &= \left(\frac{d\,V}{d\,z}\right) \qquad\qquad - \frac{d\,z^2}{d\,t^2} - \frac{\mu}{l}\,Z \end{aligned}\right\} . \quad (3)$$

Denoting by a' the distance of the point of suspension from the centre of the earth, and by ϕ' the angle between a' and the plane of the equator, we have —

$$\left.\begin{aligned} x'-x'' &= a' \cos\phi' \cos nt \\ y'-y'' &= -a' \cos\phi' \sin nt \\ z-z' &= a' \sin\phi' \end{aligned}\right\} \quad . \quad . \quad . \quad (4)$$

In differentiating the latter equations, we may consider x'', y'', z'' as constants, because the annual motion of the earth can exercise no appreciable influence upon the motion of the pendulum; we therefore have —

$$\frac{d^2x'}{dt^2} = -n^2(x'-x''); \quad \left(\frac{d^2y'}{dt^2}\right) = -n^2(y'-y''); \quad \frac{d^2z}{dt^2}=0$$

whence the equations (3) are transformed into

$$\left.\begin{aligned} \frac{d^2X}{dt} &= \left(\frac{dV}{dx}\right) - n^2X - \frac{\mu}{l}X \\ \frac{d^2Y}{dt^2} &= \frac{dV}{dy} - n^2Y - \frac{\mu}{l}Y \\ \frac{d^2Z}{dt^2} &= \left(\frac{dV}{dz}\right) \qquad - \frac{\mu}{l}Z \end{aligned}\right\} \quad . \; . \; . \; . \; . \quad (A)$$

Changing the directions of the axes of X, Y, Z into those of ζ, η, ξ, the first coinciding with the plumb-line through the point of suspension, the other directed to west, and the third to south, we have, denoting the elevation of the pole by ϕ, —

$$\left.\begin{aligned} X &= \xi \sin\phi \cos nt + \eta \sin nt - \zeta \cos\phi \cos nt \\ Y &= -\xi \sin\phi \sin nt + \eta \cos nt + \zeta \cos\phi \sin nt \\ Z &= -\xi \cos\phi \qquad\qquad - \zeta \sin\phi \end{aligned}\right\} \quad . \quad (B)$$

where

$$\xi^2 + \eta^2 + \zeta^2 = l^2.$$

Differentiating the equations (B) twice, and substituting the results into (A), we obtain, —

$$\left.\begin{aligned} &\frac{d^2\xi}{dt^2}\sin\phi\cos nt + \frac{d^2\eta}{dt^2}\sin nt - \frac{d^2\zeta}{dt^2}\cos\phi\cos nt \\ &-2n\frac{d\xi}{dt}\sin\phi\sin nt + 2n\frac{d\eta}{dt}\cos nt \\ &\qquad +2n\frac{d\zeta}{dt}\cos\phi\sin nt \end{aligned}\right\} = \left(\frac{dV}{dx}\right) - \frac{\mu}{l}X$$

$$\left.\begin{array}{l} -\frac{d^2\xi}{dt^2}\sin\phi\sin nt+\frac{d^2\eta}{dt^2}\cos nt+\frac{d^2\zeta}{dt^2}\cos\phi\sin nt \\ -2n\frac{d\xi}{dt}\sin\phi\cos nt-2n\frac{d\eta}{dt}\sin nt \\ \quad+2n\frac{d\zeta}{dt}\cos\phi\cos nt \end{array}\right\}=\left(\frac{dV}{dy}\right)-\frac{\mu}{l}Y$$

$$-\frac{d^2\xi}{dt^2}\cos\phi-\frac{d^2\zeta}{dt^2}\sin\phi \qquad =\left(\frac{dV}{dz}\right)-\frac{\mu}{l}Z.$$

Multiplying respectively by

$$\begin{array}{ccc} \sin\phi\cos nt & \sin nt & -\cos\phi\cos nt \\ -\sin\phi\sin nt & \cos nt & \cos\phi\sin nt \\ -\cos\phi & 0 & -\sin\phi \end{array}$$

then adding and considering that these factors are equal to the partial differential co-efficients

$$\begin{array}{ccc} \left(\frac{dx}{d\xi}\right) & \left(\frac{dx}{d\eta}\right) & \left(\frac{dx}{d\zeta}\right) \\ \left(\frac{dy}{d\xi}\right) & \left(\frac{dy}{d\eta}\right) & \left(\frac{dy}{d\zeta}\right) \\ \left(\frac{dz}{d\xi}\right) & \left(\frac{dz}{d\eta}\right) & .\ \left(\frac{dz}{d\zeta}\right) \end{array}$$

we obtain

$$\left.\begin{array}{l} \frac{d^2\xi}{dt^2}=-2n\frac{d\eta}{dt}\sin\phi+\left(\frac{dV}{d\xi}\right)-\frac{\mu}{l}\xi \\ \frac{d^2\eta}{dt^2}=2n\frac{d\xi}{dt}\sin\phi-2n\frac{d\zeta}{dt}\cos\phi+\left(\frac{dV}{d\eta}\right)-\frac{\mu}{l}\eta \\ \frac{d^2\zeta}{dt^2}=2n\frac{d\eta}{dt}\cos\phi+\left(\frac{dV}{d\zeta}\right)-\frac{\mu}{l}\zeta \end{array}\right\} \quad (C)$$

The assumption $\mu=0$ would lead us here to the equations of the motion of a falling or projected body, but in order to find those of the motion of the pendulum, we must eliminate this indeterminate quantity, and considering that

$$\xi\frac{d\xi}{dt}+\eta\frac{d\eta}{dt}+\zeta\frac{d\zeta}{dt}=0$$

we have,

$$\left.\begin{aligned}
\frac{\xi d^2\eta-\eta d^2\xi}{dt^2} &= 2n\cos\phi\,\xi\frac{d\zeta}{dt}-2n\sin\phi\,\zeta\frac{d\zeta}{dt}+\xi\left(\frac{dV}{d\eta}\right)\\
&\quad-\eta\left(\frac{dV}{d\xi}\right)\\
\frac{\zeta d^2\xi-\xi d^2\zeta}{dt^2} &= -2n\cos\phi\,\xi\frac{d\eta}{dt}-2n\sin\phi\,\zeta\frac{d\eta}{dt}+\zeta\left(\frac{dV}{d\xi}\right)\\
&\quad-\xi\left(\frac{dV}{d\zeta}\right)\\
\frac{\eta d^2\zeta-\zeta d^2\eta}{dt^2} &= -2n\cos\phi\,\xi\frac{d\xi}{dt}-2n\sin\phi\,\zeta\frac{d\xi}{dt}+\eta\left(\frac{dV}{d\zeta}\right)\\
&\quad-\zeta\left(\frac{dV}{d\eta}\right)
\end{aligned}\right\}\quad\text{(D)}$$

The function V consists of two parts, the one is the sum of all the particles of the earth divided by their respective distances from the attracted pendulum, the other accruing from the centrifugal force is $\frac{1}{2}\,n^2\,(x^2+y^2)$, the origin of co-ordinates being placed in the centre of the earth. On the same supposition let u, v, w denote the co-ordinates of a particle, the density of which is q, then

$$V=\int\frac{q\,du\,dv\,dw}{\sqrt{(x-u)^2+(y-v)^2+(z-w)}}+\tfrac{1}{2}n^2(x^2+y^2).$$

If the earth be considered a homogeneous ellipsoid, with semi-major axis$=a$, ellipticity$=\alpha$, and mass$=$M, denoting by r the distance of the attracted point from the centre, we would have, neglecting higher power than α^2—

$$\begin{aligned}
V=\frac{M}{r}(1-2\alpha+\alpha^2)&+\frac{Ma^2}{r^3}\left(\frac{2}{5}\alpha-\alpha^2\right)\\
&-\frac{M\,a^2z^2}{r^5}\left(\frac{3}{5}\alpha-\frac{3}{2}\alpha^2\right)+\frac{Ma^4}{r^5}\,\frac{12}{35}\,\alpha^2\\
&-\frac{M\,a^4z^2}{r^7}\,\frac{12}{7}\,\alpha^2+\frac{Ma^4z^4}{r^9}\,\frac{3}{2}\,\alpha^2+\frac{1}{2}\,n^2(x^2+y^2).
\end{aligned}$$

But if the ellipsoid is not homogeneous, and the law of its density not known, we must find the values of the coefficients in the last expression by observations. Since, for the surface of the earth, the resulting gravity

$$g = 9\overset{m}{\cdot}78019 + 0\overset{m}{\cdot}050754 \sin^2\phi$$

we may safely neglect α^2, and thus obtain

$$V = \frac{O}{r} + \frac{P}{r^3} - \frac{2\,P\,z^2}{2\,r^5} + \frac{1}{2}\,n^2\,(x^2 + y^2).$$

Taking

$$x = r\cos\vartheta\cos\vartheta'\,;\; y = r\cos\vartheta\sin\vartheta'\,;\; z = r\sin\vartheta$$

then the components of gravity will be expressed by

$$\left(\frac{dV}{dr}\right),\; \left(\frac{dV}{r\,d\vartheta}\right),\; \left(\frac{dV}{r\cos\vartheta\,d\vartheta'}\right)$$

and neglecting again α^2, we have

$$g = -\left(\frac{dV}{dr}\right) = \frac{O}{r^2} + 3\,\frac{P}{r^4}\;\frac{9\,P\sin^2\vartheta}{2\,r^4} - n^2 r\cos^2\vartheta$$

but also on the surface of the ellipsoid

$$r = a\,(1 - \alpha\sin^2\vartheta)$$

and considering that here ϕ may be written for ϑ we obtain

$$g = \frac{O}{a^2} + \frac{3\,P}{a^4} - n^2 a + \left(\frac{2\,O}{a^2}\alpha - \frac{9\,P}{2\,a^4} + n\,a^2\right)\sin^2\phi.$$

Now $\dfrac{n^2 a}{g} = \dfrac{1}{289{\cdot}24}$ and $\alpha = \dfrac{1}{300}$, whence

$$\frac{O}{a^2} = 9\overset{m}{\cdot}78183\,;\; \frac{P}{a^4} = 0{\cdot}010725\,;\; n^2 a = 0{\cdot}033813\,;$$

and thus the coefficients in the expression for V are determined.

In order to find the relations between x, y, z, and ξ, η, ζ, we must put in equations (B)

$$X = x - a' \cos \phi'; \; Y = y; \; Z = z - a' \phi'; \text{ and } n = 0$$

which gives

$$x = a' \cos \phi' + \xi \sin \phi - \zeta \cos \phi; \; y = \eta$$
$$z = a' \sin \phi' - \xi \cos \phi - \zeta \sin \phi$$

therefore, by substitution and differentiation,

$$\left(\frac{dV}{d\xi}\right) = -\left(\frac{O}{r^3} + \frac{3P}{r^5} - \frac{15}{2}\frac{Pz^2}{r^7}\right) a' \sin(\phi - \phi')$$

$$+ n^2 a' \cos \phi' \sin \phi + \frac{3P}{r^5} a' \sin \phi' \cos \phi$$

$$-\left(\frac{O}{r^3} + \frac{3P}{r^5} - \frac{15}{2}\frac{Pz^2}{r^7}\right) \xi$$

$$-\frac{3P}{r^5} \xi \cos^2\phi - \frac{3P}{r^5} \zeta \sin \phi \cos \phi + n^2 \xi \sin^2\phi$$

$$-n^2 \zeta \sin \phi \cos \phi$$

$$\left(\frac{dV}{d\eta}\right) = -\left(\frac{O}{r^3} + \frac{3P}{r^5} - \frac{15}{2}\frac{Pz^2}{r^7}\right) \eta + n^2 \eta$$

$$\left(\frac{dV}{d\zeta}\right) = \left(\frac{O}{r^3} + \frac{3P}{r^5} - \frac{15}{2}\frac{Pz^2}{r^7}\right) a' \cos(\phi - \phi')$$

$$- n^2 a' \cos \phi' \cos \phi + \frac{3P}{r^5} a' \sin \phi' \sin \phi$$

$$-\left(\frac{O}{r^3} + \frac{3P}{r^5} - \frac{15}{2}\frac{Pz^2}{r^7}\right) \zeta$$

$$-\frac{3P}{r^5} \xi \sin \phi \cos \phi - \frac{3P}{r^5} \zeta \sin^2 \phi$$

$$-n^2 \xi \sin \phi \cos \phi + n^2 \zeta \cos^2 \phi.$$

In the development of these expressions we may omit the squares and higher powers of ξ, η. ζ, since these quantities are very small compared with a'. The values for x, y z, give then,

$$r^2 = a'^2 + 2\,a'\xi \sin(\phi-\phi') - 2\,a'\zeta\cos(\phi-\phi')$$
$$z^2 = a'^2\sin^2\phi' - 2\,a'\xi\sin\phi'\cos\phi - 2\,a'\zeta\sin\phi'\sin\phi$$

therefore by substitution, and taking $\cos(\phi-\phi')$ equal to unity, we have

$$\left(\frac{dV}{d\xi}\right) = -\frac{O}{a'^2}\sin(\phi-\phi') + 3\frac{P}{a'^4}\sin\phi\cos\phi$$
$$+ n^2a'\sin\phi\cos\phi - \frac{O}{a'^3}\xi - 3\frac{O}{a'^3}\zeta\sin(\phi-\phi')$$
$$-6\frac{P}{a'^5}\xi + \frac{21}{2}\frac{P}{a'^5}\xi\sin^2\phi + 12\frac{P}{a'^5}\zeta\sin\phi\cos\phi$$
$$+ n^2\xi\sin^2\phi - n^2\zeta\sin\phi\cos\phi$$
$$\left(\frac{dV}{d\eta}\right) = -\frac{O}{a'^3}\eta - 3\frac{P}{a'^5}\eta - \frac{15}{2}\frac{P}{a'^5}\eta\sin^2\phi + n^2\eta$$
$$\left(\frac{dV}{d\zeta}\right) = \frac{O}{a'^2} + 3\frac{P}{a'^4} - \frac{9}{2}\frac{P}{a'^4}\sin^2\phi - n^2a'\cos^2\phi$$
$$+2\frac{O}{a'^3}\zeta - 3\frac{O}{a'^3}\xi\sin(\phi-\phi') + 12\frac{P}{a'^5}\zeta$$
$$-18\frac{P}{a'^5}\zeta\sin^2\phi + 12\frac{P}{a'^5}\xi\sin\phi\cos\phi$$
$$-n^2\xi\sin\phi\cos\phi + n^2\zeta\cos^2\phi$$

But referring the value of V to the point of suspension we must write

$$V = \frac{O}{a'} + \frac{P}{a'^3} - \frac{3}{2}\frac{P}{a'^3}\sin^2\phi' + \frac{1}{2}n^2a'^2\cos^2\phi'$$

and the components of gravity are then expressed by

$$\left(\frac{dV}{da'}\right) = -\frac{O}{a'^2} - 3\frac{P}{a'^4} + \frac{9}{2}\frac{P}{a'^4}\sin^2\phi' + n^2a'\cos^2\phi'$$
$$\left(\frac{dV}{a'd\phi'}\right) = -3\frac{P}{a'^4}\sin\phi'\cos\phi' - n^2a'\sin\phi'\cos\phi'$$
$$\left(\frac{dV}{a'\cos\phi' d\vartheta'}\right) = 0$$

Denoting by g' the resultant of gravity at this point we have

$$g' \cos(\phi-\phi') = -\left(\frac{dV}{da'}\right);\; g' \sin(\phi-\phi') = -\left(\frac{dV}{a'\,d\phi'}\right)$$

and therefore, with omission of the squares of ellipticity and centrifugal force,

$$g' = \frac{O}{a'^2} + 3\frac{P}{a'^4} - \frac{9}{2}\frac{P}{a'^4}\sin^2\phi - n^2 a' \cos^2\phi$$

$$0 = \frac{O}{a'^2}\sin(\phi-\phi') - 3\frac{P}{a'^4}\sin\phi\cos\phi - n^2 a' \sin\phi\cos\phi$$

whence by substitution,

$$\left(\frac{dV}{d\xi}\right) = -\frac{O}{a'^3}\xi - 6\frac{P}{a'^5}\xi + \frac{21}{2}\frac{P}{a'^5}\xi\sin^2\phi + 3\frac{P}{a'^5}\zeta\sin\phi\cos\phi$$
$$+ n^2\xi\sin^2\phi - 4n^2\zeta\sin\phi\cos\phi$$

$$\left(\frac{dV}{d\eta}\right) = -\frac{O}{a'^3}\eta - 3\frac{P}{a'^5}\eta + \frac{15}{2}\frac{P}{a'^5}\eta\sin^2\phi + n^2\eta$$

$$\left(\frac{dV}{d\zeta}\right) = g' + 2\frac{O}{a'^3},\zeta + 12\frac{P}{a'^5}\zeta - 18\frac{P}{a'^5}\zeta\sin^2\phi$$
$$+ 3\frac{P}{a'^5}\xi\sin\phi\cos\phi + n^2\zeta\cos^2\phi - 4n^2\xi\sin\phi\cos\phi.$$

For brevity's sake let us assume

$$\Gamma = \left(3\frac{P}{a'^5} + n^2\right)\cos^2\phi;\; \Lambda = \left(3\frac{P}{a'^5} - 4n^2\right)\sin\phi\cos\phi$$

$$\Pi = 3\frac{O}{a'^3} + 15\frac{P}{a'^5} - \frac{51}{2}\frac{P}{a'^5}\sin^2\phi - n^2\sin^2\phi$$

then the latter equations give by subtraction,

$$\xi\left(\frac{dV}{d\eta}\right) - \eta\left(\frac{dV}{d\xi}\right) = \Gamma\xi\eta - \Lambda\eta\zeta$$

$$\zeta\left(\frac{dV}{d\xi}\right) - \xi\left(\frac{dV}{d\zeta}\right) = -g\xi - (\Pi+\Gamma)\xi\zeta + \Lambda(\zeta^2-\xi^2)$$

$$\eta\left(\frac{dV}{d\zeta}\right) - \zeta\left(\frac{dV}{d\eta}\right) = g\eta + \Pi\eta\zeta + \Lambda\xi\eta$$

By means of these values the formulæ (D) are transformed into

$$\left.\begin{aligned}
\frac{\zeta d^2\eta-\eta d^2\zeta}{dt^2} &= -2n\cos\phi.\,\xi\frac{d\zeta}{dt}-2n\sin\phi.\,\zeta\frac{d\zeta}{dt}+\Gamma\,\xi\eta-\Lambda\,\eta\zeta \\
\frac{\zeta d^2\xi-\xi d^2\zeta}{dt^2} &= -2n\cos\phi.\,\xi\frac{d\eta}{dt}-2n\sin\phi.\,\zeta\frac{d\eta}{dt}-g\xi \\
&\quad -(\Pi+\Gamma)\,\xi\zeta+\Lambda\,(\zeta^2-\xi^2) \\
\frac{\eta d^2\zeta-\zeta d^2\eta}{dt^2} &= -2n\cos\phi.\,\xi\frac{d\xi}{dt}-2n\sin\phi.\,\zeta\frac{d\xi}{dt}+g\,\eta \\
&\quad +\Pi\,\eta\,\zeta+\Lambda\,\zeta\eta
\end{aligned}\right\}\quad (E)$$

Multiplying the second of these equations by $d\,\xi$, the third by $(-d\eta)$, adding the products and substituting $\xi\,d\,\xi+\eta\,d\,\eta=-\zeta d\,\zeta$, we get after integration,

$$\frac{d\,\xi^2+d\,\eta^2+d\,\zeta^2}{d\,t^2}=\kappa+2\,g\,\zeta+\Pi\,\zeta^2-\Gamma\,\xi^2+2\,\Lambda\,\xi\,\zeta \quad . \quad . \quad (F)$$

where κ denotes the arbitrary constant. This formula, which is the only complete integral obtainable from (E), expresses the relative velocity of the simple pendulum in term of the co-ordinates of its material point.

The Compound Pendulum.—The differential equations of the motion of the compound pendulum, which can turn freely about its point of suspension, are obtained by multiplying the preceding equations by $(d\,m)$, and then integrating with respect to this differential, and for the whole extent of the mass of the pendulum.

Let us refer all the particles of the vibrating body to the three rectangular axes u, v, w, which intersect at the point of suspension, and are at the same time principal axes of the body, the centre of gravity of which is assumed to lie within the axis of w. Denoting the angle between the axes w and ζ by θ; further, the angle between the planes $w\,\zeta$ and $\xi\,\zeta$ by ψ; finally, that between the planes $w\,\zeta$ and $u\,w$ by χ, we have, with introduction of auxiliary co-ordinates,

$$\xi=\xi' \cos\psi-\eta' \sin\psi;\ \xi'=\ \ \xi'' \cos\theta+\zeta'' \sin\theta$$
$$\xi''=u \cos\chi-v \sin\chi$$
$$\eta=\xi' \sin\psi+\eta' \cos\psi;\ \eta'=\eta''$$
$$\eta''=u \sin\chi+v \cos\chi$$
$$\zeta=\zeta';\ \zeta'=-\xi'' \sin\theta+\zeta'' \cos\theta$$
$$\zeta''=w$$

whence, by elimination,

$$\xi=\alpha\, u+\beta\, v+\gamma\, w;\ \eta=\alpha'\, u+\beta' v+\gamma'\, w$$
$$\zeta=\alpha'' u+\beta'' v+\gamma'' w. \quad \ldots\ldots\ldots \quad \text{(G)}$$

where

$$\alpha=-\sin\chi \sin\psi+\cos\chi \cos\psi \cos\theta$$
$$\beta=-\cos\chi \sin\psi-\sin\chi \cos\psi \cos\theta$$
$$\gamma=\cos\psi \sin\theta$$
$$\alpha'=\ \ \sin\chi \cos\psi+\cos\chi \sin\psi \cos\theta$$
$$\beta'=\ \ \cos\chi \cos\psi-\sin\chi \sin\psi \cos\theta$$
$$\gamma'=\sin\psi \sin\theta$$
$$\alpha''=-\cos\chi \sin\theta$$
$$\beta''=\sin\chi \sin\theta$$
$$\gamma''=\cos\theta$$

and the equations of condition between these quantities are

$$\alpha^2+\ \beta^2+\ \gamma^2=\ \alpha'^2+\ \beta'^2+\ \gamma'^2=\ \alpha''^2+\ \beta''^2+\ \gamma''^2=1$$
$$\alpha\alpha'+\beta\beta'+\gamma\gamma'=\alpha\alpha''+\beta\beta''+\gamma\gamma''=\alpha'\alpha''+\beta'\beta''+\gamma'\gamma''=0$$

or,

$$\alpha^2+\alpha'^2+\ \ \alpha''^2=\beta^2+\ \beta'^2+\ \ \beta''^2=\gamma^2+\gamma'^2+\gamma''^2\ =1$$
$$\alpha\beta+\alpha'\beta'+\alpha''\beta''=\alpha\gamma+\alpha'\gamma'+\alpha''\gamma''=\beta\gamma+\beta'\gamma'+\beta''\gamma''=0$$

from which we may derive

$$\alpha\beta' - \alpha'\beta = \gamma'' \,;\ \alpha''\beta - \alpha\beta'' = \gamma' \,;\ \alpha'\beta'' - \alpha''\beta' = \gamma$$
$$\alpha'\gamma - \alpha\gamma' = \beta'' \,;\ \alpha\gamma'' - \alpha''\gamma = \beta' \,;\ \alpha''\gamma' - \alpha'\gamma'' = \beta$$
$$\beta\gamma' - \beta'\gamma = \alpha'' \,;\ \beta''\gamma - \beta\gamma'' = \alpha' \,;\ \beta'\gamma'' - \beta''\gamma' = \alpha$$

Denoting the elements of rotation about the axes u, v, w, by $p\,dt$, $q\,dt$, $r\,dt$, respectively, we have,

$$r\,dt = d\,\chi + \cos\theta\,d\,\psi$$
$$q\,dt = \sin\chi\sin\theta\,d\psi + \cos\chi\,d\theta$$
$$p\,dt = -\cos\chi\sin\theta\,d\psi + \sin\chi\,d\theta$$

and the differentiation of the preceding values of α, β, &c., gives

$$d\alpha = (\beta r - \gamma q)\,dt\,;\ d\beta = (\gamma p - \alpha r)\,dt$$
$$d\gamma = (\alpha q - \beta p)\,dt$$
$$d\alpha' = (\beta' r - \gamma' q)\,dt\,;\ d\beta' = (\gamma' p - \alpha' r)\,dt$$
$$d\gamma' = (\alpha' q - \beta' p)\,dt$$
$$d\alpha'' = (\beta'' r - \gamma'' q)\,dt\,;\ d\beta'' = (\gamma'' p - \alpha'' r)\,dt$$
$$d\gamma'' = (\alpha'' q - \beta'' p)\,dt$$

whence, differentiating the three equations of (G) we obtain,

$$\left.\begin{aligned} d\xi &= \alpha\,(wq - vr)\,dt + \beta\,(ur - wp)\,dt \\ &\quad + \gamma\,(vp - uq)\,dt \\ d\eta &= \alpha'\,(wq - vr)\,dt + \beta'\,(ur - wp)\,dt \\ &\quad + \gamma'\,(vp - uq)\,dt \\ d\zeta &= \alpha''\,(wq - vr)\,dt + \beta''\,(ur - wp)\,dt \\ &\quad + \gamma''\,(vp - uq)\,dt \end{aligned}\right\} \quad . \quad . \quad (H)$$

Since u, v, w are principal axes of the pendulum, we have,

$$\int uv\,dm = 0,\ \int uw\,dm = 0,\ \int vw\,dm = 0$$

and if A, B, C denote the momentums of inertia in respect to the same axes, we have

$$A=\int (v^2+w^2)\, dm,\ B=\int (u^2+w^2)\, dm,\ C=\int (u^2+v^2)\, dm$$

and the foregoing formulæ become, by means of the equations of condition,

$$\left.\begin{aligned} \int (\xi d\eta-\eta\, d\,\xi)\, dm&=(A\alpha'' p+B\beta'' q+C\gamma'' r)\, dt \\ \int (\zeta d\xi-\xi\, d\,\zeta)\, dm&=(A\alpha' p+B\beta' q+C\gamma' r)\, dt \\ \int (\eta d\zeta-\zeta\, d\,\eta)\, dm&=(A\alpha\ p+B\beta\ q+C\gamma\ r)\, dt \end{aligned}\right\} \quad . \quad . \quad (I)$$

and by squaring the equations (H), and then adding,

$$\int (d\xi^2+d\eta^2+d\zeta^2)\, dm=(Ap^2+Bq^2+Cr^2)\, dt^2 \quad . \quad . \quad (K)$$

Differentiating the three equations (H) in respect to the time, we obtain

$$\begin{aligned} \int (\xi d^2\eta-\eta d^2\xi)\, dm=&(A\alpha''\, dp+B\beta''\, dq+C\gamma''\, dr)\, dt \\ &+[(C-B)\,\alpha''\, qr+(A-C)\,\beta''\, pr \\ &+(B-A)\,\gamma''\, pq]dt^2 \\ \int (\zeta d^2\xi-\xi d^2\zeta)\, dm=&(A\alpha'\, dp+B\beta'\, dq+C\gamma'\, dr)\, dt \\ &+[(C-B)\,\alpha'\, qr+(A-C)\,\beta'\, pr \\ &+(B-A)\,\gamma'\, pq]\, dt^2 \\ \int (\eta d^2\zeta-\zeta d^2\eta)\, dm=&(A\alpha\, dp+B\beta\, dq+C\gamma\, dr)\, dt \\ &+[(C-B)\,\alpha\, qr+(A-C)\,\beta\, pr \\ &+(B-A)\,\gamma\, pq]\, dt^2 \end{aligned}$$

whence we derive

$$\begin{aligned} C\, dr\ dt+(B-A)\ pq\ dt^2=&\int [\gamma\ (\eta d^2\zeta-\zeta d^2\eta) \\ &+\gamma'\ (\zeta d^2\xi-\xi d^2\zeta)+\gamma''\ (\xi d^2\eta-\eta d^2\xi)]\ dm \quad . \quad . \quad (L) \end{aligned}$$

and this equation is transformed by means of the formulæ (D) into

$$C\frac{dr}{dt}+(B-A)pq=\int\Big[g(\gamma\eta-\gamma'\xi)-2n\cos\phi\Big(\gamma\frac{d\xi}{dt}+\gamma'\frac{d\eta}{dt}+\gamma''\frac{d\zeta}{dt}\Big)\xi$$
$$-2n\sin\phi\Big(\gamma\frac{d\xi}{dt}+\gamma'\frac{d\eta}{dt}+\gamma''\frac{d\zeta}{dt}\Big)\zeta$$
$$+\Pi(\gamma\eta-\gamma'\xi)\zeta+\Gamma(\gamma''\eta-\gamma'\zeta)\xi$$
$$+\Lambda\left[(\gamma'\zeta-\gamma''\eta)\zeta+(\gamma\eta-\gamma'\xi)\xi\right]\Big]dm$$

The equation (H) gives

$$\gamma d\xi+\gamma' d\eta+\gamma'' d\zeta=(vp-uq)\,dt$$

from which we obtain by means of (G)

$$\int(\gamma d\xi+\gamma' d\eta+\gamma'' d\zeta)\,\xi\,dm=-\tfrac{1}{2}\,C\,d\gamma-\tfrac{1}{2}(B-A)\ (\alpha q+\beta p)\,dt$$

$$\int(\gamma d\xi+\gamma' d\eta+\gamma'' d\zeta)\,\zeta\,dm=-\tfrac{1}{2}\,C\,d\gamma''-\tfrac{1}{2}(B-A)(\alpha'' q+\beta'' p)\,dt.$$

The formulæ (G) give further in conjunction with the equations of conditions

$$\gamma\eta-\gamma'\xi=\beta'' u-\alpha'' v;\ \gamma''\eta-\gamma'\zeta=\alpha v-\beta u$$

whence

$$\int(\gamma\eta-\gamma'\xi)\,\zeta\,dm=\alpha''\beta''(B-A)$$
$$\int(\gamma''\eta-\gamma'\zeta)\,\xi\,dm=-\alpha\beta\,(B-A)$$
$$\int(\gamma\eta-\gamma'\xi)\,\xi dm=-\tfrac{1}{2}\gamma'\,C+\tfrac{1}{2}(\alpha\beta''+\alpha''\beta)(B-A)$$
$$\int(\gamma'\zeta-\gamma''\eta)\,\zeta\,dm=\tfrac{1}{2}\gamma'\,C+\tfrac{1}{2}(\alpha\beta''+\alpha''\beta)(B-A).$$

Since the axis of w passes through the centre of gravity, we have denoting the distance of the latter from the point of suspension by λ,—

$$\int u\,dm=0;\int v\,dm=0;\int w\,dm=\lambda m$$

and therefore,

$$\int(\gamma\eta-\gamma'\xi)\,dm=0.$$

Substituting these integrals, we obtain

$$C\frac{dr}{dt}+(B-A)\,pq=Cn\cos\phi\,\frac{dr}{dt}+Cn\sin\phi\,\frac{d\gamma''}{dt}$$
$$+(B-A)\,n\cos\phi\,(\alpha q+\beta p)$$
$$+(B-A)\,n\sin\phi\,(\alpha''q+\beta''p)$$
$$+(B-A)\;\Pi\alpha''\beta''-(B-A)\;\Gamma\alpha\beta$$
$$+(B-A)\,\Lambda\,(\alpha\beta''+\alpha''\beta);$$

but in all well-constructed pendulums we will have two momentums of inertia equal to each other. Assuming, therefore, as is generally the case, that

$$A=B,$$

we have, by integration,

$$r=n'+\gamma n\cos\phi+\gamma''n\sin\phi$$
$$=n'+n\cos\phi\cos\psi\sin\theta+n\sin\phi\cos\theta \quad (A_1)$$

where n' is the arbitrary constant, and expresses the angular velocity which was originally communicated to the pendulum about the axis that passes through its centre of gravity and point of suspension.

The equations (F) and (K) give the following expression:

$$Ap^2+Bq^2+Cr^2=\kappa+2g\int\zeta\,dm+\Pi\int\zeta^2\,dm$$
$$-\Gamma\int\xi^2\,dm+2\,\Lambda\int\xi\zeta\,dm.$$

But

$$\int\xi^2\,dm=-\alpha^2A-\beta^2B-\gamma^2\,C+\tfrac{1}{2}\,(A+B+C)$$

$$\int\zeta^2\,dm=\quad\alpha''^2A-\beta''^2\,B-\gamma''^2\,C+\tfrac{1}{2}\,(A+B+C)$$

$$\int\xi\zeta\,dm=-\alpha\alpha''A-\beta\beta''B-\gamma\gamma''C$$

$$\int\zeta\,dm=\lambda m\gamma'',$$

therefore, denoting by κ the sum of all constants

$$Ap^2+Bq^2+Cr^2=\kappa+2\,g\lambda\,m\gamma''-\Pi\,(\alpha''^2\,A+\beta''^2B+\gamma''^2C)$$
$$+\Gamma\,(\alpha^2A+\beta^2B+\gamma^2C)$$
$$-2\,\Lambda\,(\alpha\alpha''\,A+\beta\beta''B+\gamma\gamma''\,C).$$

Substituting the value of r^2 from (A_1), taking $B=A$, and observing that $p^2+q^2=\frac{d\theta^2}{dt^2}+\frac{d\psi^2}{dt^2}\sin^2\theta$, while $\gamma=\cos\psi\sin\theta$; and $\gamma''=\cos\theta$, the last equation becomes

$$A\,\frac{d\theta^2+d\psi^2\sin^2\theta}{dt^2}=\kappa+2\,(g\lambda m-Cnn'\sin\phi)\cos\theta$$
$$-2\,Cnn'\cos\phi\cos\psi\sin\theta$$
$$+\ [\Pi\,(A-C)-Cn^2\sin^2\phi]\cos^2\theta$$
$$+2\,[\Lambda\,(A-C)-Cn^2\sin\phi\cos\phi]\cos\psi\sin\theta\cos\theta$$
$$-\ [\Gamma\,(A-C)+Cn^2\cos^2\phi]\cos^2\psi\sin^2\theta. \quad . \quad . \quad (B_1).$$

This integral, in conjunction with (A_1), shows that the conservation of vis viva takes place.

We further have by means of the first equations of (E) and (I),—

$$d\,(A\alpha''p+B\beta''q+C\gamma''r)=-2n\,\cos\phi\int\xi d\zeta\,dm$$
$$-2n\sin\phi\int\zeta d\zeta\,dm$$
$$+dt\,\Gamma\int\xi\eta\,dm-dt\,\Lambda\int\eta\zeta\,dm.$$

But

$$\int\xi d\zeta\,dm=-\alpha d\alpha''A-\beta d\beta''B-\gamma d\gamma''C$$
$$+\tfrac{1}{2}\,(\alpha d\alpha''+\beta d\beta''+\gamma d\gamma'')\,(A+B+C)$$
$$\int\zeta d\,\zeta dm=-\alpha''\,d\alpha''\,A-\beta''d\beta''\,B-\gamma''d\,\gamma''C$$
$$\int\xi\eta\,dm=-\alpha\alpha'A-\beta\beta'B-\gamma\gamma'C$$
$$\int\eta\zeta\,dm=-\alpha'\alpha''\,A-\beta'\beta''B-\gamma'\,\gamma''C\cdot$$

Therefore, by substitution and integration :

$$A\alpha''p+B\beta''q+C\gamma''r=c+n\sin\phi\,(\alpha''^2A+\beta''^2B+\gamma''^2C)$$

$$+2n\cos\phi\int(\alpha d\alpha''A+\beta d\beta''B+\gamma d\gamma''C)$$

$$-n\cos\phi\,(A+B+C)\int(\alpha d\alpha''+\beta d\beta''+\gamma d\gamma'')$$

$$-\Gamma\int(\alpha\alpha'A+\beta\beta'B+\gamma\gamma'C)\,dt$$

$$+\Lambda\int(\alpha'\alpha''A+\beta'\beta''B\times\gamma'\gamma''C)\,dt.$$

Taking, again, A=B, substituting the value of r from (A_1), and observing that

$$d\alpha''=(\beta''r-\gamma''q)\,dt\,;\;d\beta''=(\gamma''p-\alpha''r)dt\,;\text{ etc.}$$

$$\alpha''p+\beta''q=\frac{d\psi}{dt}\sin^2\theta\,;\;\gamma'=\sin\psi\sin\theta\,;\text{ etc.}$$

we obtain finally,

$$A\frac{d\psi}{dt}\sin^2\theta=c-An\sin\phi\cos^2\theta-Cn'\cos\theta$$

$$+2\,An\,\cos\phi\int\cos\psi\sin^2\theta d\theta$$

$$+Cnn'\cos\phi\int\sin\psi\sin\theta dt$$

$$-[\Lambda(A-C)-Cn^2\sin\phi\cos\phi]\int\sin\phi\cos\theta\sin\theta\,dt$$

$$+[\Gamma(A-C)+Cn^2\cos^2\phi]\int\sin\psi\cos\psi\sin^2\theta dt\,.\,.\quad(C_1)$$

This integral shows that the conservation of areas does not take place.

The further integration of (B_1) and (C_1) can only be effected by a series of approximations; the first of which will suffice to show the laws of the motion of the free pendulum. For this purpose we take the following terms from the equation (B_1) and (C_1),—

$$A\left(\sin^2\theta\frac{d\psi^2}{dt^2}+\frac{d\theta^2}{dt^2}\right)=\kappa+2(g\lambda m-\mathrm{Cnn'}\sin\phi)\cos\theta$$

$$A\sin^2\theta\frac{d\psi}{dt}=c-\mathrm{An}\sin\phi\cos^2\theta-\mathrm{Cn'}\cos\theta.$$

Assuming for shortness

$$\mu=n\sin\phi;\ \mu'=n'\frac{C}{A};\ g'=g\frac{\lambda m}{A}$$

we have

$$A\left(\sin^2\theta\frac{d\psi^2}{dt^2}+\frac{d\theta^2}{dt^2}\right)=\kappa+2\,A\,(g'-\mu\mu')\cos\theta\ .\ \ .\ (D_1)$$

$$A\sin^2\theta\frac{d\psi}{dt}=c-A\,\mu\cos^2\theta-A\,\mu'\cos\theta\quad .\ (E_1)$$

Denoting by ε the angle which the pendulum made with the plumb line at the moment of liberation, and by v the velocity of the lateral impulse communicated in a direction perpendicular to the primitive plane of oscillation, then we have at the commencement of motion,

$$\theta=\varepsilon;\ \frac{d\theta}{dt}=0;\ \frac{d\psi}{dt}=v$$

We find, therefore, by substitution the values of the constants,

$$\kappa=A\,v^2\sin^2\varepsilon-2\,A\,(g'-\mu\mu')\cos\varepsilon$$
$$c=A\,v\ \sin^2\varepsilon+A\,\mu\cos^2\varepsilon+A\,\mu'\cos\varepsilon$$

and the equations (D_1) and (E_1) are transformed into

$$\sin^2\theta\frac{d\psi^2}{dt^2}+\frac{d\theta^2}{dt^2}+2\,(g'-\mu\mu')(\cos\varepsilon-\cos\theta)=v^2\sin^2\varepsilon$$

$$\sin^2\theta\frac{d\psi}{dt}=\mu\,(\cos^2\varepsilon-\cos^2\theta)+\mu'\,(\cos\varepsilon-\cos\theta)+v\sin^2\varepsilon$$

where v and ε are the arbitrary constants which we will eliminate by introducing two others κ and κ' determined by the equations.

$$\left.\begin{aligned} \sin^2\theta \frac{d\psi^2}{dt^2} + \frac{d\theta^2}{dt^2} + 2(g'-\mu\mu')(1-\cos\theta) &= \kappa \\ \sin^2\theta \frac{d\psi}{dt} = \mu \sin^2\theta + \mu'(1-\cos\theta) &+ \kappa' \end{aligned}\right\} \quad . \quad (F_1)$$

whence, by comparing these with the preceding equations, we have,—

$$\left.\begin{aligned} \kappa &= 2(g'-\mu\mu')(1-\cos\varepsilon)+v^2\sin^2\varepsilon \\ \kappa' &= -\mu\sin^2\varepsilon - \mu'(1-\cos\varepsilon)+v\sin^2\varepsilon \end{aligned}\right\} \quad . \quad . \quad . \quad (G_1)$$

These equations show that κ and κ' are small quantities of the second order, if ε is a small quantity of the first order as will be assumed hereafter agreeable to practice. Taking the value of $\frac{d\psi^2}{dt^2}$ from the second equation (F_1), and substituting it into the first, we obtain,—

$$0 = \sin^2\theta \frac{d\theta^2}{dt^2} + 2g'\sin^2\theta(1-\cos\theta)+\mu^2\sin^4\theta - \kappa''\sin^2\theta$$
$$+\mu'^2(1-\cos\theta)^2+2\kappa'\mu'(1-\cos\theta)+\kappa'^2$$

where for shortness

$$\kappa'' = \kappa - 2\kappa'\mu.$$

Let now

$$x = 1-\cos\theta$$

and the preceding equation becomes —

$$0 = \frac{dx^2}{dt^2} + \kappa'^2 - 2\,\mathrm{D}x + \mathrm{E}x^2 - 2\,\mathrm{F}x^3 + \mu^2 x^4 \quad . \quad . \quad (H_1)$$

where

$$\begin{aligned} \mathrm{D} &= \kappa'' - \kappa'\mu' \\ \mathrm{E} &= 4g' + \kappa'' + 4\mu^2 + \mu'^2 \\ \mathrm{F} &= g' + 2\mu^2. \end{aligned}$$

Since we assume the initial value of θ, that is ε, to be a small quantity, we may be satisfied with an approximate integration by means of a converging series. For this purpose we differentiate a second time, thus —

$$0 = \frac{d^2x}{dt^2} - D + E\,x - 3\,F\,x^2 + 2\mu^2 x^3$$

where E and F are finite quantities, but D a small quantity of the second order, and the coefficient of x^3 at least of the fourth order, as we will see hereafter, We assume, therefore, the integral to have the following form,—

$$x = a + b \cos 2\,(\omega\,t + \eta) + c \cos 4\,(\omega\,t + \eta)$$
$$+ e \cos 6\,(\omega\,t + \eta) + ;\ \text{etc.}$$

where a and b are small quantities of the second order, c of the fourth, e of the sixth order, and so on.

Differentiating and substituting this value of x in the last differential equation, we have for the determination of the coefficients the following equations,—

$$0 = -D + E\,a - 3\,F\,(a^2 + \tfrac{1}{2}\,b^2)$$
$$0 = -4\,\omega^2 + E - 3\,F\,(2\,a + c)$$
$$0 = -16\,\omega^2\,c + E\,c - 3\,F\,(2\,ac + \tfrac{1}{2}b^2)$$
$$0 = -36\,\omega^2\,e + E\,e - 3\,F\,bc.$$

These equations leave b and η arbitrary, but since $\frac{dx}{dt} = 0$, when $t = 0$ we have

$$\eta = 0$$

and since at the same time $x = 1 - \cos\varepsilon$, we have

$$1 - \cos\varepsilon = a + b + c + e + ;\ \text{etc.}$$

As it is sufficient for our purpose only to consider the small quantities of the second order, we take

$$x = a + b \cos 2\,\omega\,t$$

where the coefficients are to be determined by the equations

$$a+b=\tfrac{1}{2}\varepsilon^2;\ a=\frac{D}{E};\ 4\,\omega^2=E-6\,F\,a=E-6\,\frac{DF}{E}.$$

The equations (G_1) become, not considering smaller quantities than those of the second order,

$$\kappa=(v^2+g'-\mu\mu')\,\varepsilon^2$$
$$\kappa'=\tfrac{1}{2}\,(2\,v-2\,\mu-\mu')\,\varepsilon^2.$$

Substituting the values of D and E we obtain, neglecting ε^2,

$$a=\frac{2\,v^2+2\,g'-2\,\mu\mu'-(2\,\mu+\mu')(2\,v-2\,\mu-\mu')}{4\,g'+4\,\mu^2+\mu'^2}\cdot\frac{\varepsilon^2}{2}$$

which, combined with

$$a+b=\tfrac{1}{2}\,\varepsilon^2$$

gives

$$a-b=\frac{(2\,v-2\,\mu-\mu')^2}{4\,g'+4\,\mu^2+\mu'^2}\cdot\frac{\varepsilon^2}{2}=\frac{1}{2}\varepsilon'^2$$

where

$$\varepsilon'=\frac{2\,v-2\mu-\mu'}{\sqrt{4g'+4\mu^2+\mu'^2}}\,\varepsilon.$$

But from $x=a+b\cos 2\,\omega\,t$ we derive, considering that $x=1-\cos\theta=2\sin^2\tfrac{1}{2}\,\theta$, and interchanging sine and arc,—

$$x=\tfrac{1}{2}\,\theta^2=(a+b)\cos^2\omega\,t+(a-b)\sin^2\omega\,t$$

that is,

$$\theta^2=\varepsilon^2\cos^2\omega+\varepsilon'^2\sin^2\omega\,t \quad . \quad . \quad (A_2)$$

Whence we see that while ε expresses the greatest, ε' expresses the least deviation of the pendulum from the plumb line.

Substituting the values of D, E, and F in the equation for ω^2, we get,—

$$\omega^2 = g' + \mu^2 + \frac{1}{4}\mu'^2 + \frac{1}{4}\kappa'' - \frac{3}{8}\frac{(g' + +2\mu^2)(\kappa'' - \kappa'\mu')}{g' + \mu^2 + \frac{1}{4}\mu'^2} \quad (B_2)$$

and by neglecting the quantities smaller than those of the second order,—

$$\varepsilon' = \frac{2v - 2\mu - \mu'}{2\omega}\varepsilon \quad . \quad . \quad . \quad . \quad . \quad (C_2)$$

whence

$$\kappa' = \omega\,\varepsilon\,\varepsilon' \quad . \quad . \quad . \quad . \quad . \quad . \quad (D_2)$$

Putting this value of κ' in the second equation (F_1), and neglecting again quantities of higher than the second order, we have

$$d\psi = (\mu + \frac{1}{2}\mu')\,dt + \frac{\omega\,\varepsilon\,\varepsilon'}{\theta^2}\,dt$$

and substituting for θ^2 its value from equation (A_2),

$$d\psi = (\mu + \frac{1}{2}\mu')\,dt + \frac{\omega\,\varepsilon\,\varepsilon'\,dt}{\varepsilon^2\cos^2\omega t + \varepsilon'^2\sin^2\omega t}$$

But $\displaystyle\int \frac{\omega\,\varepsilon\,\varepsilon'\,dt}{\varepsilon^2\cos^2\omega t + \varepsilon'^2\sin^2\omega t} = \frac{\varepsilon'}{\varepsilon}\int \frac{\omega\,dt}{\cos^2\omega t\,(1 + \frac{\varepsilon'^2}{\varepsilon^2}\tan^2\omega t)}$

$$= \frac{\varepsilon'}{\varepsilon}\int \frac{d\tan\omega t}{1 + \frac{\varepsilon'^2}{\varepsilon^2}\tan^2\omega t} = \tan^{-1}(\frac{\varepsilon'}{\varepsilon}\tan\omega t)$$

Therefore,

$$\psi = \alpha + (\mu + \frac{1}{2}\mu')\,t + \tan^{-1}(\frac{\varepsilon'}{\varepsilon}\tan\omega t) \quad . \quad (E_2)$$

The arbitrary constant α denotes the azimuth of the plane in which the pendulum commenced to swing.

The equation shows that the plane of oscillation changes and moves in consequence of the rotation of the earth from east through south to west except μ' be negative, or greater than 2μ. Let us find the curve of intersection of the moving pendulum, with a horizontal plane in the distance = 1 from the point of suspension. Taking in this plane the axis of x positive towards south, and the axis of y positive towards west, we obtain for the co-ordinates of the pendulum the following expressions :—

$$x=\theta \cos \psi \qquad\qquad y=\theta \sin \psi$$

but the preceding integral gives

$$\tan (\psi - \mathrm{M}) = \frac{\varepsilon'}{\varepsilon} \tan \omega t \quad . \quad . \quad . \quad . \quad (\mathrm{F}_2)$$

if we assume

$$\mathrm{M} = \alpha + (\mu + \frac{1}{2} \mu') t$$

hence

$$\cos^2 \omega t = \frac{\varepsilon'^2 \cos^2 (\psi - \mathrm{M})}{\varepsilon'^2 \cos^2 (\psi - \mathrm{M}) + \varepsilon^2 \sin^2 (\psi - \mathrm{M})}$$

$$\sin^2 \omega t = \frac{\varepsilon^2 \sin^2 (\psi - \mathrm{M})}{\varepsilon'^2 \cos^2 (\psi - \mathrm{M}) + \varepsilon^2 \sin^2 (\psi - \mathrm{M})}$$

and substituting these values in equation (A_2) we obtain

$$\frac{\theta^2 \cos^2 (\psi - \mathrm{M})}{\varepsilon^2} + \frac{\theta^2 \sin^2 (\psi - \mathrm{M})}{\varepsilon'^2} = 1.$$

This is the equation to an ellipse, the co-ordinates of which are

$$x' = \theta \cos (\psi - \mathrm{M}); \qquad y' = \theta \sin (\psi - \mathrm{M})$$

and its principal axes are $2\,\varepsilon$ and $2\,\varepsilon'$. Comparing these co-ordinates with those of x and y above, we find

$$x = x' \cos \mathrm{M} - y' \sin \mathrm{M}$$
$$y = x' \sin \mathrm{M} + y' \cos \mathrm{M}$$

whence we infer that the axes of this ellipse move from east to west, with an angular velocity expressed by

$$(\mu + \frac{1}{2} \mu').$$

The pendulum, therefore, moves on the surface of a right cone, with elliptical basis, and the axes of this basis change continually with the stated velocity their position with regard to the meridian of the place of observation.

The magnitude of the minor axis $2\,\varepsilon'$ depends chiefly on the velocity v of the lateral impulse which we assumed the pendulum might receive at the commencement of its motion.

If

$$2\,v-2\,\mu-\mu'=2\,\omega$$

then the base of the cone on which the pendulum moves is circular.

The change of the plane of oscillation of the pendulum in the azimuth amounts during each mean second of time, in consequence of the earth's rotation, to—

$$\mu=(15''\cdot041)\sin\phi$$

We must now examine the amount of the velocity of this change in consequence of the rotation of the pendulum on its own axis.

We must first remark that since μ is always positive on the northern hemisphere, and μ' either positive or negative, according as the rotation of the pendulum on its own axis takes place from east to west or from west to east, the two changes in azimuth will either conspire or oppose each other.

Let us denote by T the duration of oscillation between two consecutive greatest elongations of the pendulum, then the change in azimuth after each oscillation is expressed by

$$(\mu+\frac{1}{2}\,\mu')\,\mathrm{T}.$$

Let us assume, as is generally the case, that the bob consists of a homogeneous ball, the diameter of which is small compared with the length of the suspending wire, while the mass of the latter is very small compared with that of the former; then

$$A=\lambda^2 m$$

and

$$C=\frac{2}{5}\rho^2 m$$

where ρ denotes the radius of the ball of the pendulum, and substituting this value in the expression for μ', we have,—

$$\mu'=\frac{2\,\rho^2}{5\,\lambda^2}\,n'.$$

Let the number of rotations of the pendulum on its axis during each oscillation be denoted by u, then

$$n'\,T=2\,u\,\pi.$$

and sufficiently approximate

$$T=\pi\sqrt{\frac{\lambda}{g}}$$

whence

$$\mu'=\frac{4\,u\,\rho^2}{5\,\lambda^2}\sqrt{\frac{g}{\lambda}}$$

therefore,

$$\frac{1}{2}\,\mu'\,T=\left(\frac{2\,\pi}{5}\right)\left(\frac{\rho}{\lambda}\right)^2 u=72^\circ\left(\frac{\rho}{\lambda}\right)^2 u.$$

Taking $\lambda=40\,\rho$, as nearly was the case in Foucault's experiments, then

$$\frac{1}{2}\,\mu'\,T=162''\,u.$$

Whence we see that a moderate velocity of rotation of

the pendulum on its own axis, caused by a lateral impulse, may effect a change in the azimuth equal to, or even greater than that in consequence of the earth's rotation, and either in the same or contrary direction.

This change in the azimuth does not depend on the absolute length of the pendulum, but on the ratio of the latter to the radius of the bob, which for this reason ought to consist of a matter of great specific gravity. We also infer from the last formula how necessary it is to avoid a rotation of the pendulum on its own axis.

Let us suppose that this rotation is scarcely appreciable and that at the same time $\varepsilon=\varepsilon'$ very nearly, i. e., that the pendulum is moving on the surface of a cone with nearly circular base.

Taking in this case

$$\gamma=\varepsilon-\varepsilon'$$

then γ is, compared with ε and ε', a very small quantity, the square and higher powers of which may be neglected.

The equation (F_2) gives therefore, by making $\mu'=0$

$$\tan(\psi-\alpha-\mu t)=\tan\omega t-\frac{\gamma}{\varepsilon}\tan\omega t \quad . \; . \; . \quad (G_2)$$

from which we derive, neglecting the square of γ,

$$\psi=\alpha+(\mu+\omega)t-\frac{\gamma}{2\varepsilon}\sin 2\omega t.$$

Since the value of μ is always positive on the northern, and negative on the southern hemisphere of the Earth, it follows that the time of an entire revolution of ψ is different according as ω is positive or negative, that is, according as the pendulum moves from east through south to west, or from west through south to east. On the northern hemisphere of the Earth, ψ will pass through the whole circumference in shorter time, if the motion of the pendulum takes place in the first direction; the contrary will be the case on the southern hemisphere.

The equation (G_2) gives $\psi-\alpha=0$ for $t=0$; hence taking $\psi-\alpha=\pm 2i\pi$, according as ω is positive or negative, we obtain for the determination of the time during which ψ passes i times the whole circumference, the following equation—

$$\pm 2i\pi=(\omega+\mu)t-\frac{\gamma}{2\varepsilon}\sin 2\omega t.$$

Taking approximately

$$t=\frac{\pm 2i\pi}{\omega+\mu}$$

we have

$$2\omega t=\pm 2i\pi\mp 2i\pi\frac{\mu}{\omega}$$

therefore

$$\sin 2\omega t=\mp 2i\pi\frac{\mu}{\omega}$$

and by substituting this value in the first equation

$$\pm 2i\pi=(\omega+\mu)t\pm\frac{i\pi\gamma\mu}{\varepsilon\omega}$$

whence we derive, neglecting the square of μ,

$$t=\pm 2i\pi\frac{1-\frac{\mu\gamma}{2\varepsilon\omega}}{\omega+\mu}=\pm\frac{2i\pi}{\omega}\left(1-\frac{\mu}{\omega}-\frac{\gamma\mu}{2\varepsilon\omega}\right).$$

The difference of the times of revolution of ψ arising from the difference of the direction in which the pendulum moves, is therefore

$$\frac{4\pi\mu}{\omega^2}\left(1+\frac{\gamma}{2\varepsilon}\right)=\frac{4\pi}{13714\,.\,g}\left(1+\frac{\gamma}{2\varepsilon}\right)\lambda\sin\phi.$$

We infer therefore that this difference is not depending on the primitive amplitude of oscillation, if the motion of the pendulum takes place on the surface of a cone with circular base, where $\gamma=0$, and furthermore that this difference is but very little changed if the base differs little from a circle, because the quantity $\frac{\gamma}{2\varepsilon}$ is in this case very small.

Expressed in seconds of time, we have, since

$$\log \frac{4\pi}{13714 \,.\, g}=5{,}9705-10$$

and taking $\lambda=10$ metres, $\phi=51°$, and $\gamma=0$, the following numerical value for the difference of the times of a revolution of ψ

$$0'',00073.$$

This is, indeed, a small fraction of time, yet it may be ascertained by observation in well-conducted experiments, as Mr. Bravais has shown. (See p. 105.)

In order to find expressions for $\int r dt$ and χ, we take the equation (A_1), neglecting the square and higher powers of θ; thus we get

$$r=n'+\mu+n \cos \phi \,.\, \theta \cos \psi.$$

But the equation (F_2) gives

$$\theta \sin (\psi-M)=\varepsilon' \sin \omega t$$
$$\theta \cos (\psi-M)=\varepsilon \cos \omega t$$

whence we obtain

$$\theta \cos \psi=\tfrac{1}{2}(\varepsilon+\varepsilon') \cos (\omega t+M)+\tfrac{1}{2}(\varepsilon-\varepsilon') \cos (\omega t-M).$$

Substituting this value of $\theta \cos \psi$, multiplying by dt and integrating, we have

$$\int r\,dt = \beta + (n' + \mu)t + \frac{(\varepsilon + \varepsilon')\,n \cos\phi}{2\,\omega} \sin(\omega t + \mathrm{M})$$
$$+ \frac{(\varepsilon - \varepsilon')\,n \cos\phi}{2\,\omega} \sin(\omega t - \mathrm{M})$$

where we have neglected in the denominator the quantity $(\mu + \frac{1}{2}\mu')$, and where β denotes the arbitrary constant. But

$$\int r dt = \chi + \psi$$

hence, by means of equation (E_2),

$$\chi = \beta - \alpha + (n' - \tfrac{1}{2}\mu')t - \tan^{-1}\left(\frac{\varepsilon'}{\varepsilon} \tan \omega t\right) + \frac{(\varepsilon + \varepsilon')n \cos\phi}{2\,\omega} \sin(\omega t + \mathrm{M})$$
$$+ \frac{(\varepsilon - \varepsilon')\,n \cos\phi}{2\,\omega} \sin(\omega t - \mathrm{M}).$$

In this equation the effect of the Earth's rotation on the change of the plane of oscillation in the azimuth is eliminated; from the same expression we derive the values which χ assumes at the greatest amplitudes, viz:—

$$\chi = \beta - \alpha + (n' - \tfrac{1}{2}\mu')i\mathrm{T} + i\pi + \frac{\varepsilon' n \cos\phi}{\omega} \sin\left[\alpha + (\mu + \tfrac{1}{2}\mu')i\mathrm{T}\right.$$

the coefficient of the last term of this expression is very small, never exceeding one second, moreover it is $=0$, if $\varepsilon' = 0$, and may therefore be neglected, hence

$$\chi = \beta - \alpha + i\,\pi + (n' - \tfrac{1}{2}\mu')\,i\mathrm{T}$$
$$= \beta - \alpha + i\,\pi + n' \left(1 - \frac{\mathrm{C}}{2\mathrm{A}}\right) i\mathrm{T}$$

Assuming that the pendulum received no rotation on its own axis at the commencement of its motion, we have the coefficient of T equal to zero, and therefore

$$\chi = \beta - \alpha + i\,\pi$$

that is, at any two consecutive greatest amplitudes, the values of χ are respectively $\beta-\alpha$ and $180°+\beta-\alpha$; hence the following theorem.

If we at the commencement of the motion of the pendulum draw through its axis and the vertical line a plane which remains fixed relatively to the pendulum, then this plane will, notwithstanding the motion of the plane of oscillation in the azimuth, always pass through the vertical line at every greatest amplitude, except some extraneous force communicated to the pendulum at the commencement of its motion a rotation on its own axis.

Further approximation to the integration of the equations for the motion of the compound pendulum.

In order to effect the following approximations to the integration of our differential equations, we will make use of the method for the variation of arbitrary constants.

The equation $(F_{,})$ to be integrated may be written:

$$\frac{d\,\theta^2}{d\,t^2}=-\sin^2\theta\,\frac{d\,\psi^2}{d\,t^2}-2(g'-\mu\mu')\,(1-\cos\theta)+\kappa$$

$$\sin^2\theta\,\frac{d\,\psi}{d\,t}=\mu\sin^2\theta+\mu'(1-\cos\theta)+\kappa'$$

where, abstracting from the correction for κ which is not wanted in this case,

$$\kappa=\Pi\cos^2\theta+2\,\Lambda\cos\psi\sin\theta\cos\theta-\Gamma\cos^2\psi\sin^2\theta$$
$$-2\,\frac{\mathrm{C}}{\mathrm{A}}\,n\,n'\cos\phi\cos\psi\sin\theta$$

$$\frac{d\,\kappa'}{d\,t}=\tfrac{1}{2}\left(\frac{d\,\kappa}{d\,\psi}\right)+2\,n\cos\phi\,\cos\psi\,\sin^2\theta\,\frac{d\,\theta}{d\,t}.$$

By differentiation we obtain

$$2\frac{d\theta}{dt}\frac{d^2\theta}{dt^2}=-2\sin^2\theta\frac{d\psi}{dt}\frac{d^2\psi}{dt^2}-2\sin\theta\cos\theta\frac{d\theta}{dt}\frac{d\psi^2}{dt^2}$$
$$-2\sin\theta\,(g'-\mu\mu')\frac{d\theta}{dt}+\frac{d\kappa}{dt}$$

$$\sin^2\theta\frac{d^2\psi}{dt^2}=-2\sin\theta\cos\theta\frac{d\theta}{dt}\frac{d\psi}{dt}+2\mu\sin\theta\cos\theta\frac{d\theta}{dt}$$
$$+\mu'\sin\theta\frac{d\theta}{dt}+\frac{d\kappa'}{dt}.$$

Eliminating $\frac{d^2\psi}{dt^2}$ from the first by means of the second equation, we have

$$2\frac{d\theta}{dt}\frac{d^2\theta}{dt^2}=2\sin\theta\cos\theta\frac{d\theta}{dt}\frac{d\psi^2}{dt^2}-2(g'-\mu\mu')\sin\theta\frac{d\theta}{dt}$$
$$-4\mu\sin\theta\cos\theta\frac{d\theta}{dt}\frac{d\psi}{dt}-2\mu'\sin\theta\frac{d\theta}{dt}\frac{d\psi}{dt}+\frac{d\kappa}{dt}-2\frac{d\psi}{dt}\frac{d\kappa'}{dt}.$$

But

$$d\psi=\mu+\mu'\frac{1-\cos\theta}{\sin^2\theta}+\frac{\kappa'}{\sin^2\theta}$$

and for very small arcs of oscillation

$$g'=\omega^2-\mu^2-\tfrac{1}{4}\mu'^2;\quad \kappa'=\omega\varepsilon\varepsilon'$$

hence by substitution

$$\frac{d^2\theta}{dt^2}=-\omega^2\sin\theta+\mu^2\sin\theta\,(1-\cos\theta)-\mu'^2\frac{\sin^3\frac{1}{2}\theta\,(1+\cos^2\frac{1}{2}\theta)}{2\cos^3\frac{1}{2}\theta}$$
$$-\mu'\omega\varepsilon\varepsilon'\frac{\sin\frac{1}{2}\theta}{2\cos^3\frac{1}{2}\theta}+\omega^2\varepsilon^2\varepsilon'^2\frac{\cos\theta}{\sin^3\theta}+\frac{\frac{d\kappa}{dt}}{2\frac{d\theta}{dt}}-\frac{\frac{d\psi}{dt}\cdot\frac{d\kappa'}{dt}}{\frac{d\theta}{dt}}$$

$$\frac{d^2\psi}{dt^2}=-\frac{d\theta}{dt}\left(2\omega\varepsilon\varepsilon'\frac{\cos\theta}{\sin^3\theta}-\mu'\frac{\sin\frac{1}{2}\theta}{2\cos^3\frac{1}{2}\theta}\right)+\frac{1}{\sin^2\theta}\cdot\frac{d\kappa'}{dt}.$$

The integrals with which these expressions must be compared, are the equation (A_2) and (E_2) after introducing in them the arbitrary constant η, viz.

$$\theta^2=\varepsilon^2\cos^2(\omega t+\eta)+\varepsilon'^2\sin^2(\omega t+\eta)$$
$$\psi=\alpha+(\mu+\tfrac{1}{2}\mu')t+\tan^{-1}\left(\frac{\varepsilon'}{\varepsilon}\tan(\omega t+\eta)\right)$$

where ε, ε', η and α denote the four arbitrary constants. If these constants are considered as not variable, then these equations express the approximate integrals of the preceding differential equations of the second order, and they become the complete integrals if the four constants are taken variable and their variations suitably determined. Differentiating the integrals and taking that part of the differential which arises from the variation of the arbitrary constants equal to zero, we obtain

$$\theta\frac{d\theta}{dt} = -\omega(\varepsilon^2-\varepsilon'^2)\sin(\omega t+\eta)\cos(\omega t+\eta)$$

$$\frac{d\psi}{dt} = \mu + \tfrac{1}{2}\mu' + \frac{\omega\varepsilon\varepsilon'}{\theta^2}$$

$$0 = \frac{d\varepsilon}{dt}\varepsilon\cos^2(\omega t+\eta) + \frac{d\varepsilon'}{dt}\varepsilon'\sin^2(\omega t+\eta) - \frac{d\eta}{dt}(\varepsilon^2-\varepsilon'^2)\sin(\omega t+\eta)\cos(\omega t+\eta)$$

$$0 = \theta^2\frac{d\alpha}{dt} - \frac{d\varepsilon}{dt}\varepsilon'\sin(\omega t+\eta)\cos(\omega t+\eta) + \frac{d\varepsilon'}{dt}\varepsilon\sin(\omega t+\eta)\cos(\omega t+\eta) + \frac{d\eta}{dt}\varepsilon\varepsilon'$$

Differentiating a second time and considering that

$$\frac{1}{\theta}\frac{d\theta^2}{dt^2} = \frac{\omega^2}{\theta^3}(\varepsilon^2-\varepsilon'^2)^2\sin^2(\omega t+\eta)\cos^2(\omega t+\eta)$$
$$= \omega^2\theta - \frac{\omega^2\varepsilon^2\varepsilon'^2}{\theta^3} - \frac{\omega^2(\varepsilon^2-\varepsilon'^2)}{\theta}\left[\cos^2(\omega t+\eta)-\sin^2(\omega t+\eta)\right]$$

we obtain

$$\frac{d^2\theta}{dt^2} = -\omega^2\theta + \frac{\omega^2\varepsilon^2\varepsilon'^2}{\theta^3} - 2\frac{d\varepsilon}{dt}\frac{\omega\varepsilon\sin(\omega t+\eta)\cos(\omega t+\eta)}{\theta} + 2\frac{d\varepsilon'}{dt}\frac{\omega\varepsilon'\sin(\omega t+\eta)\cos(\omega t+\eta)}{\theta} - \frac{d\eta}{dt}\frac{\omega(\varepsilon^2-\varepsilon'^2)}{\theta}\left[\cos^2(\omega t+\eta)-\sin^2(\omega t+\eta)\right]$$

$$\frac{d^2\psi}{dt^2} = -\frac{2\,\omega\,\epsilon\,\epsilon'}{\theta^3}\frac{d\theta}{dt} - \frac{d\epsilon}{dt}\frac{\omega\,\epsilon'}{\theta^4}\left[\epsilon^2\cos^2(\omega t+\eta) - \epsilon'^2\sin^2(\omega t+\eta)\right]$$

$$+\frac{d\epsilon'}{dt}\frac{\omega\,\epsilon}{\theta^4}\left[\epsilon^2\cos^2(\omega t+\eta) - \epsilon'^2\sin^2(\omega t+\eta)\right]$$

$$+\,2\frac{d\eta}{dt}\frac{\omega\,\epsilon\,\epsilon'}{\theta^4}(\epsilon^2-\epsilon'^2)\sin(\omega t+\eta)\cos(\omega t+\eta).$$

Comparing these values of $d^2\theta$ and $d^2\psi$ with those found before, and taking

$$\mathrm{D} = \omega^2\theta^2 - \frac{\omega^2\varepsilon^2\varepsilon'^2}{\theta^2} - \omega^2\theta\sin\theta + \mu^2\theta\sin\theta\,(1-\cos\theta)$$

$$-\mu'^2\theta\,\frac{\sin^3\frac{1}{2}\theta(1+\cos^2\frac{1}{2}\theta)}{2\cos^3\frac{1}{2}\theta} - \mu'\omega\varepsilon\varepsilon'\theta\,\frac{\sin\frac{1}{2}\theta}{2\cos^3\frac{1}{2}\theta}$$

$$+\,\omega^2\varepsilon^2\varepsilon'^2\frac{\theta\cos\theta}{\sin^3\theta} + \frac{\theta\frac{d\kappa}{dt}}{2\frac{d\theta}{dt}} - \frac{\theta\frac{d\psi}{dt}\frac{d\kappa'}{dt}}{\frac{d\theta}{dt}}$$

$$\mathrm{E} = \frac{2\,\omega\varepsilon\varepsilon'}{\theta}\frac{d\theta}{dt} - \theta^2\frac{d\theta}{dt}\left(2\,\omega\varepsilon\varepsilon'\frac{\cos\theta}{\sin^3\theta} - \mu'\frac{\sin\frac{1}{2}\theta}{2\cos^3\frac{1}{2}\theta}\right) + \frac{\theta^2}{\sin^2\theta}\frac{d\kappa'}{dt}$$

we obtain the following equations

$$\mathrm{D} = -2\frac{d\epsilon}{dt}\,\omega\,\epsilon\sin(\omega t+\eta)\cos(\omega t+\eta) + 2\frac{d\epsilon'}{dt}\,\omega\,\epsilon'\sin(\omega t+\eta)\cos(\omega t+\eta)$$

$$-\frac{d\eta}{dt}\,\omega\,(\epsilon^2-\epsilon'^2)\left[\cos^2(\omega t+\eta) - \sin^2(\omega t+\eta)\right]$$

$$\theta^2\,\mathrm{E} = -\frac{d\epsilon}{dt}\,\omega\,\epsilon'\left[\epsilon^2\cos^2(\omega t+\eta) - \epsilon'^2\sin^2(\omega t+\eta)\right]$$

$$+\frac{d\epsilon'}{dt}\,\omega\,\epsilon\left[\epsilon^2\cos^2(\omega t+\eta) - \epsilon'^2\sin^2(\omega t+\eta)\right]$$

$$+\,2\,\frac{d\eta}{dt}\,\omega\,\epsilon\,\epsilon'\,(\epsilon^2-\epsilon'^2)\sin(\omega t+\eta)\cos(\omega t+\eta).$$

These equations in conjunction with the two others, viz.,

$$\frac{d\epsilon}{dt}\,\epsilon\cos^2(\omega t+\eta) + \frac{d\epsilon'}{dt}\,\epsilon'\sin^2(\omega t+\eta) - \frac{d\eta}{dt}\,(\epsilon^2-\epsilon'^2)\sin(\omega t+\eta)\cos(\omega t+\eta) = 0$$

$$\theta^2\frac{d\alpha}{dt} - \frac{d\epsilon}{dt}\,\epsilon'\sin(\omega t+\eta)\cos(\omega t+\eta) + \frac{d\epsilon'}{dt}\,\epsilon\sin(\omega t+\eta)\cos(\omega t+\eta) + \frac{d\eta}{dt}\,\epsilon\,\epsilon' = 0$$

are sufficient to determine the differentials of the constants which are considered to become variable.

We derive from them the following equations,

$$\frac{d\varepsilon}{dt} = -\frac{D\varepsilon}{\omega\theta^2}\sin(\omega t+\eta)\cos(\omega t+\eta) + \frac{E\varepsilon'}{\omega\theta^2}\sin^2(\omega t+\eta)$$

$$\frac{d\varepsilon'}{dt} = \frac{D\varepsilon'}{\omega\theta^2}\sin(\omega t+\eta)\cos(\omega t+\eta) + \frac{E\varepsilon}{\omega\theta^2}\cos^2(\omega t+\eta)$$

$$\frac{d\eta}{dt} = -\frac{D}{\omega\theta^2(\varepsilon^2-\varepsilon'^2)}[\varepsilon^2\cos^2(\omega t+\eta) - \varepsilon'^2\sin^2(\omega t+\eta)]$$
$$+\frac{2E\varepsilon\varepsilon'}{\omega\theta^2(\varepsilon^2-\varepsilon'^2)}\sin(\omega t+\eta)\cos(\omega t+\eta)$$

$$\frac{d\alpha}{dt} = \frac{D\varepsilon\varepsilon'}{\omega\theta^2(\varepsilon^2-\varepsilon'^2)}[\cos^2(\omega t+\eta) - \sin^2(\omega t+\eta)]$$
$$-\frac{E(\varepsilon^2+\varepsilon'^2)}{\omega\theta^2(\varepsilon^2-\varepsilon'^2)}\sin(\omega t+\eta)\cos(\omega t+\eta).$$

We see that the quantity θ^2 occurs in the denominators of all the terms. If now the motion of the pendulum is such that $\varepsilon'=0$, then likewise $\theta=0$ at every oscillation when the pendulum coincides with the plumbline, consequently the values of the variations of $\varepsilon, \varepsilon', \eta$ and α would become infinitely great unless this factor θ is expunged by the substitution of the values of D and E containing the same factor in the numerator. Since we know by observation that the motion of the pendulum is continuous, we may conclude that all the terms of D and E must be so composed, that after substituting the values of the latter in the preceding equations, the factor θ^2 is generated in every numerator, by which means the same factor is eliminated from the denominators. It will be seen hereafter that this is really the case.

Furthermore, the factor $\varepsilon^2-\varepsilon'^2$ in the denominator of the expressions for $d\eta$ and $d\alpha$ shows that the variations of these two constants become infinitely great if $\varepsilon=\varepsilon'$. In this case the arbitrary constants ε, ε', η and α, become inapplicable, and we must, therefore, make use of others. Let us denote the new constants by β, h, l, α', taking

$$\beta=\tfrac{1}{2}(\varepsilon+\varepsilon');\quad h=\tfrac{1}{2}(\varepsilon-\varepsilon')\sin 2\eta$$
$$\alpha'=\alpha+\eta;\quad l=\tfrac{1}{2}(\varepsilon-\varepsilon')\cos 2\eta$$

whence

$$d\beta=\tfrac{1}{2}(d\varepsilon+d\varepsilon');\, dh=\tfrac{1}{2}(d\varepsilon-d\varepsilon')\sin 2\eta+d\eta\,(\varepsilon-\varepsilon')\cos 2\eta$$
$$d\alpha'=d\alpha+d\eta;\quad dl=\tfrac{1}{2}(d\varepsilon-d\varepsilon')\cos 2\eta-d\eta(\varepsilon-\varepsilon')\sin 2\eta.$$

Dividing these equations by dt, and substituting the values of the differentials of ε, ε', α and η, we obtain

$$\frac{d\beta}{dt}=-\frac{D}{2\omega\theta^2}(l\sin 2\omega t+h\cos 2\omega t)+\frac{E}{2\omega\theta^2}(\beta+l\cos 2\omega t-h\sin 2\omega t)$$
$$\frac{dh}{dt}=-\frac{D}{2\omega\theta^2\beta}(\beta^2\cos 2\omega t+2\beta l-hl\sin 2\omega t+l^2\cos 2\omega t)$$
$$+\frac{E}{2\omega\theta^2\beta}(\beta^2\sin 2\omega t-\beta h-hl\cos 2\omega t-l^2\sin 2\omega t)$$
$$\frac{dl}{dt}=-\frac{D}{2\omega\theta^2\beta}(\beta^2\sin 2\omega t-2\beta h-hl\cos 2\omega t+h^2\sin 2\omega t)$$
$$-\frac{E}{2\omega\theta^2\beta}(\beta^2\cos 2\omega t+\beta l-hl\sin 2\omega t-h^2\cos 2\omega t)$$
$$\frac{d\alpha'}{dt}=-\frac{D}{2\omega\theta^2\beta}(\beta+l\cos 2\omega t-h\sin 2\omega t)-\frac{E}{2\omega\theta^2\beta}(l\sin 2\omega t+h\cos 2\omega t)$$

where the divisor $(\varepsilon-\varepsilon')$ has disappeared. It remains to be shown that θ and ψ may be expressed in functions of the new constants. For shortness we put

$$\beta'=\tfrac{1}{2}(\varepsilon-\varepsilon')=\sqrt{h^2+l^2}$$

then the equation

$$\theta^2=\varepsilon^2\cos^2(\omega t+\eta)+\varepsilon'^2\sin^2(\omega t+\eta)$$

gives by substitution the following

$$\theta^2=\beta^2+\beta'^2+2\beta\,(l\cos 2\omega t-h\sin 2\omega t)$$

by means of which, after substituting for β' its value, the quantity θ will be expressed in functions of β, h, and l. From the equation (F_2) we derive

$$\sin(\psi - M - \omega t - \eta) = -\frac{2\beta'}{\beta+\beta'}\sin(\omega t+\eta)\cos(\psi+M)$$

but the same equation gives

$$\cos(\psi - M) = \frac{\sqrt{\beta+\beta'}\,.\,\cos(\omega t+\eta)}{\sqrt{\beta + l\cos 2\omega t - h\sin 2\omega t}}$$

whence by substitution

$$\psi = (\omega+\mu+\tfrac{1}{2}\mu')\,t + \alpha' - \sin^{-1}\frac{l\sin 2\omega t + h\cos 2\omega t}{\sqrt{\beta+\beta'}\,.\,\sqrt{\beta + l\cos 2\omega t - h\sin 2\omega t}}$$

by means of which, since $\beta' = \sqrt{h^2+l^2}$, the quantity ψ is expressed in functions of β, h, l and α'.

In order to effect the elimination of D and E, we must successively substitute the several parts of D and E. Let us first take that part of D and E which is not depending on the differentials of κ and κ', retaining in each expression only the first not-vanishing term; then because

$$\frac{\cos\theta}{\sin^3\theta} = \frac{1}{\theta^3} - 15\,\theta + \text{etc.}$$

we have

$$D = \theta^4\left(\frac{1}{6}\omega^2 + \frac{1}{2}\mu^2 - \frac{1}{8}\mu'^2\right) - \theta^2\omega\varepsilon\varepsilon'\left(\frac{1}{4}\mu' + \frac{1}{15}\omega\varepsilon\varepsilon'\right)$$

$$E = \theta^3\frac{d\theta}{dt}\left(\frac{1}{4}\mu' + \frac{2}{15}\omega\varepsilon\varepsilon'\right).$$

In these expressions we may consider as vanishing μ and μ' with respect to ω, and likewise the terms of the sixth order in comparison to those of the fourth order; thus we obtain

$$D = \frac{1}{6}\omega^2\theta^4;\quad E = 0.$$

Substituting these values and that of θ^2 into those of the differentials of ε, ε', η and α, we get

$$\frac{d\epsilon}{dt}=-\frac{1}{6}\omega\,[\epsilon^3\cos^3(\omega t+\eta)\sin(\omega t+\eta)+\epsilon\,\epsilon'^2\sin^3(\omega t+\eta)\cos(\omega t+\eta)]$$

$$\frac{d\epsilon'}{dt}=\frac{1}{6}\omega\,[\epsilon^2\epsilon'\cos^3(\omega t+\eta)\sin(\omega t+\eta)+\epsilon'^3\sin^3(\omega t+\eta)\cos(\omega t+\eta)]$$

$$\frac{d\eta}{dt}=-\frac{1}{16}\omega\,(\epsilon^2+\epsilon'^2)-\frac{\omega\,(\epsilon^4+\epsilon'^4)}{12\,(\epsilon^2-\epsilon'^2)}\cos 2\,(\omega t+\eta)-\frac{\omega}{48}(\epsilon^2+\epsilon'^2)\cos 4\,(\omega t+\eta)$$

$$\frac{d\alpha}{dt}=\frac{1}{24}\omega\,\epsilon\,\epsilon'+\frac{\omega\,(\epsilon^2+\epsilon'^2)}{12\,(\epsilon^2-\epsilon'^2)}\epsilon\,\epsilon'\cos 2\,(\omega t+\eta)+\frac{\omega}{24}\epsilon\,\epsilon'\cos 4\,(\omega t+\eta)$$

Integrating and denoting the increment of the constants by a prefixed δ, we have

$$\delta\epsilon=\frac{1}{24}\epsilon^3\cos^4(\omega t+\eta)-\frac{1}{24}\epsilon\,\epsilon'^2\sin^4(\omega t+\eta)$$

$$\delta\epsilon'=-\frac{1}{24}\epsilon^2\epsilon'\cos^4(\omega t+\eta)+\frac{1}{24}\epsilon'^3\sin^4(\omega t+\eta)$$

$$\delta\eta=-\frac{1}{16}\omega\,(\epsilon^2+\epsilon'^2)\,t-\frac{1}{24}\frac{\epsilon^4+\epsilon'^4}{\epsilon^2-\epsilon'^2}\sin 2\,(\omega t+\eta)-\frac{1}{192}(\epsilon^2+\epsilon'^2)\sin 4\,(\omega t+\eta)$$

$$\delta\alpha=\frac{1}{24}\omega\,\epsilon\,\epsilon'\,t+\frac{1}{24}\frac{\epsilon^2+\epsilon'^2}{\epsilon^2-\epsilon'^2}\epsilon\,\epsilon'\sin 2\,(\omega t+\eta)+\frac{1}{96}\epsilon\,\epsilon'\sin 4\,(\omega t+\eta)$$

where $\eta=0$ on the right-hand side of the equations. The terms which are affected with the time, t, require particular notice, but the other terms are so small that they modify the motion of the pendulum very little, though they hold good for every case, even if $\varepsilon=\varepsilon'$, as will be seen by substituting them in the equation to the base of the conical surface on which the pendulum moves, the factor $(\varepsilon^2-\varepsilon'^2)$ being destroyed by this substitution.

The term in the value of $\delta\eta$ which is multiplied by t shows that, instead of the arc ωt of the first approximation, we must substitute $[\omega-\frac{1}{16}\omega\,(\varepsilon^2+\varepsilon'^2)]t$, where $\omega=\sqrt{g'+\mu^2+\frac{1}{4}\mu'^2}$, whence we obtain for the time of oscillation

$$\mathrm{T}=\frac{\pi}{\sqrt{g'+\mu^2+\frac{1}{4}\mu'^2}}\left(1+\frac{\varepsilon^2+\varepsilon'^2}{16}\right).$$

The term in the value of $\delta\alpha$ which is multiplied by t, shows that the motion of the plane of oscillation, or, which amounts to the same, the motion of the great axis of the elliptical base of the conical surface on which the pendulum moves, is expressed by

$$\left(\mu+\frac{1}{2}\mu'+\frac{1}{24}\omega\varepsilon\varepsilon'\right)t$$

instead of by $(\mu+\frac{1}{2}\mu')t$ as found in the first approximation. The additive correction which this expression has received by the more approximate integration of our differential equations may under circumstances become very considerable.

Taking the equation (C_2) viz—

$$\varepsilon'=\frac{2\upsilon-2\mu-\mu'}{2\,\omega}\varepsilon$$

referring it to the initial values of ε and ε', and assuming the initial lateral impulse communicated to the pendulum, $\upsilon=0$, we would have

$$\varepsilon'=-\frac{2\,\mu+\mu'}{2\,\omega}\varepsilon$$

and therefore the correction for the motion of the plane of oscillation is

$$-\frac{1}{24}\left(\mu+\frac{1}{2}\mu'\right)\varepsilon^2 t.$$

This value is even for the greatest initial amplitudes generally allowed more than a thousand times smaller than $(\mu+\frac{1}{2}\mu')t$, and consequently of little effect. But if the condition $\upsilon=0$ is not completely fulfilled, the correction may become considerable. It will be always possible to ascertain the initial lateral velocity by means of the value of ε' which can be observed as well as that of ε, and these two quantities in conjunction with the length of the pendulum will furnish the value of the correction.

Let us illustrate by a numerical example how great the latter may become in certain cases.

Suppose the length of the pendulum

$$\lambda = 10^m$$

then since very nearly

$$\omega = \sqrt{\frac{g}{\lambda}};\ g = 9^m, 8$$

we may assume

$$\omega = 1.$$

Let, as has been the case in several experiments of this kind,

$$\varepsilon = \frac{1}{9};\ \varepsilon' = \frac{1}{400}$$

then we obtain

$$\frac{1}{24}\,\omega\,\varepsilon\,\varepsilon' = 2'', 4$$

but for the middle of Europe we have nearly

$$\mu = 11''$$

so that the effect of the correction amounts almost to $\frac{1}{4}$ of the value of the principal term, and much smaller values of ε' than the one here assumed may exert an appreciable influence upon the motion of the plane of oscillation.

Hence it follows that in all experiments with the free pendulum it is necessary to observe not alone the greatest but also the least deviation of the pendulum from the plumb line in order that we may be able to compare the results of observation with those of theory.

The corrections here found exert on the difference of the times of revolution of ψ, if $\varepsilon = \varepsilon'$, only an inappreciable influence. For by means of them we get, since ε and ε' are always positive, and taking $\mu'=0$, the following expression for ψ

$$\psi=\alpha+\mu t+\omega\left(1-\frac{\varepsilon^2}{12}\right)t$$

whence we obtain for the difference of the times of revolution of ψ which takes place according as the pendulum moves from left to right, or from right to left the expression

$$\frac{4\pi\mu}{\omega^2}\left(1+\frac{\varepsilon^2}{6}\right)$$

so that, if for instance $\varepsilon=\frac{1}{9}$, the numerical value given in the first approximation is only increased by its 486th part.

We now proceed to investigate those parts of D and E which depend on the differentials of κ and κ'. Having only regard to these parts, and taking $\sin\theta=\theta$, we obtain

$$D=\frac{\theta\frac{d\kappa}{dt}}{2\frac{d\theta}{dt}}-\frac{\theta\frac{d\psi}{dt}\frac{d\kappa'}{dt}}{\frac{d\theta}{dt}};\quad E=\frac{d\kappa'}{dt}.$$

But κ is a function of the variables ψ and θ, and therefore

$$\frac{d\kappa}{dt}=\left(\frac{d\kappa}{d\psi}\right)\frac{d\psi}{dt}+\left(\frac{d\kappa}{d\theta}\right)\frac{d\theta}{dt}$$

further sufficiently exact

$$\frac{d\kappa'}{dt}=\tfrac{1}{2}\left(\frac{d\kappa}{d\psi}\right)+2n\cos\phi\cos\psi\,.\,\theta^2\frac{d\theta}{dt}.$$

whence by substitution

$$D-\tfrac{1}{2}\theta\left(\frac{d\kappa}{d\theta}\right)-2n\cos\phi\cos\psi\,.\,\theta^3\frac{d\psi}{dt}$$

$$E=\tfrac{1}{2}\left(\frac{d\kappa}{d\psi}\right)+2n\cos\phi\cos\psi\,.\,\theta^2\frac{d\Gamma}{dt}$$

by means of which the differential equations of ϵ, ϵ', η and α are transformed into the following:

$$\left.\begin{array}{l}
\frac{d\epsilon}{dt} = \frac{\sin(\omega t+\eta)}{2\omega\theta^2}\left[\left(\frac{d\kappa}{d\psi}\right)\epsilon'\sin(\omega t+\eta) - \theta\left(\frac{d\kappa}{d\theta}\right)\epsilon\cos(\omega t+\eta)\right] \\
\qquad + \frac{2n\cos\phi}{\omega}\cos\psi\sin(\omega t+\eta)\left[\theta\frac{d\psi}{dt}\epsilon\cos(\omega t+\eta) + \frac{d\theta}{dt}\epsilon'\sin(\omega t+\eta)\right] \\
\frac{\delta\epsilon'}{dt} \quad \frac{\cos(\omega t+\eta)}{2\omega\theta^2}\left[\left(\frac{d\kappa}{d\psi}\right)\epsilon\cos(\omega t+\eta) + \theta\left(\frac{d\kappa}{d\theta}\right)\epsilon'\sin(\omega t+\eta)\right] \\
\qquad \frac{2n\cos\phi}{\omega}\cos\psi\cos(\omega t+\eta)\left[\theta\frac{d\psi}{dt}\epsilon'\sin(\omega t+\eta) - \frac{d\theta}{dt}\epsilon\cos(\omega t+\eta)\right] \\
\frac{d\eta}{dt} \quad - \frac{1}{(\epsilon^2-\epsilon'^2)}\left[\left(\frac{d\kappa}{d\psi}\right)\epsilon\epsilon'\sin(\omega t+\eta)\cos(\omega t+\eta)\right. \\
\qquad\qquad \left. - \frac{1}{2}\theta\left(\frac{d\kappa}{d\theta}\right)\left(\epsilon^2\cos^2(\omega t+\eta) - \epsilon'^2\sin^2(\omega t+\eta)\right)\right] \\
\qquad + \frac{2n\cos\phi}{\omega(\epsilon^2-\epsilon'^2)}\cos\psi\left[\theta\frac{d\psi}{dt}\left(\epsilon^2\cos^2(\omega t+\eta) - \epsilon'^2\sin^2(\omega t+\eta)\right)\right. \\
\qquad\qquad \left. + 2\frac{d\theta}{dt}\epsilon\epsilon'\sin(\omega t+\eta)\cos(\omega t+\eta)\right] \\
\frac{d\alpha}{dt} = -\frac{1}{2\omega\theta^2(\epsilon^2-\epsilon'^2)}\left[\left(\frac{d\kappa}{d\psi}\right)(\epsilon^2+\epsilon'^2)\sin(\omega t+\eta)\cos(\omega t+\eta)\right. \\
\qquad\qquad \left. - \theta\left(\frac{d\kappa}{d\theta}\right)\epsilon\epsilon'\left(\cos^2(\omega t+\eta) - \sin^2(\omega t+\eta)\right)\right] \\
\qquad - \frac{2n\cos\phi}{\omega(\epsilon^2-\epsilon'^2)}\cos\psi\left[\theta\frac{d\psi}{dt}\epsilon\epsilon'\left(\cos^2(\omega t+\eta) - \sin^2(\omega t+\eta)\right)\right. \\
\qquad\qquad \left. + \frac{d\theta}{dt}(\epsilon^2+\epsilon'^2)\sin(\omega t+\eta)\cos(\omega t+\eta)\right]
\end{array}\right\} (a)$$

We will first consider the terms arising from $\left(\frac{d\kappa}{d\psi}\right)$ and $\left(\frac{d\kappa}{d\theta}\right)$. Neglecting $\frac{\text{C}}{\text{A}}$ as small compared to A, we have

$$\kappa = \Pi\cos^2\theta + 2\Lambda\cos\psi\sin\theta\cos\theta - \Gamma\cos^2\psi\sin^2\theta$$

whence by differentiation and taking $\sin\theta = \theta$

$$\left(\frac{d\kappa}{d\psi}\right) = -2\Lambda\,\theta\sin\psi + 2\Gamma\,\theta\sin\psi\,.\,\theta\cos\psi$$

$$\theta\left(\frac{d\kappa}{d\theta}\right) = -2\,\Pi\,\theta^2 + 2\,\Lambda\,\theta\cos\psi - 4\,\Lambda\,\theta^2\,.\,\theta\cos\psi - 2\,\Gamma\,\theta\cos\psi\,.\,\theta\cos\psi.$$

In these formulas the term affected with Π is much greater than the others, and requires therefore the first consideration. We substitute therefore in the differential equation of ε, ε', η and α

$$\left(\frac{d\kappa}{d\psi}\right)=0;\ \theta\left(\frac{d\kappa}{d\theta}\right)=-2\,\Pi\,\theta^2$$

making on the right hand side of these equation $\eta=0$; thus we get

$$\frac{d\varepsilon}{dt}=\frac{\Pi\,\varepsilon}{2\,\omega}\sin 2\,\omega t$$

$$\frac{d\varepsilon'}{dt}=-\frac{\Pi\,\varepsilon'}{2\,\omega}\sin 2\,\omega t$$

$$\frac{d\eta}{dt}=\frac{\Pi}{2\,\omega}+\frac{\Pi\,(\varepsilon^2+\varepsilon'^2)}{2\,\omega(\varepsilon^2-\varepsilon'^2)}\cos 2\,\omega t$$

$$\frac{d\alpha}{dt}=\frac{\Pi\,\varepsilon\,\varepsilon'}{\omega(\varepsilon^2-\varepsilon'^2)}\cos 2\,\omega t$$

Integrating these equations so that the variations of the elements (arbitrary constants) become zero for $t=0$, and denoting them by a prefixed δ, then

$$\delta\,\varepsilon=\frac{\Pi\,\varepsilon}{4\,\omega^2}(1-\cos 2\,\omega\,t)$$

$$\delta\varepsilon'=-\frac{\Pi\,\varepsilon'}{4\,\omega^2}(1-\cos 2\,\omega t)$$

$$\delta\eta=\frac{\Pi}{2\,\omega}t+\frac{\Pi\,(\varepsilon^2+\varepsilon'^2)}{4\,\omega\,(\varepsilon^2-\varepsilon'^2)}\sin 2\,\omega\,t$$

$$\delta\alpha=\frac{\Pi\,\varepsilon\,\varepsilon'}{2\,\omega^2\,(\varepsilon^2-\varepsilon'^2)}\sin 2\,\omega\,t.$$

The extreme smallness of the recurring terms in these equations shows that the curvature of the earth, considered as a sphere, has no appreciable influence on the change of the plane of oscillation in the azimuth, as might have been seen à priori. The expression for $\delta\eta$ shows that we must substitute in the value for the time of oscillation $\left(\omega + \frac{\Pi}{2\,\omega}\right)$ for (ω); thus we get

$$T = \frac{\pi}{\omega}\left(1 - \frac{\Pi}{2\,\omega^2}\right).$$

Taking again the approximate value

$$\omega^2 = \frac{g}{\lambda}$$

the factor for the correction of the time of oscillation arising from the spherical figure of the earth, is

$$1 - \frac{\Pi\,\lambda}{2\,g}.$$

Now considering only the greatest term in the value of Π, we have (see page 169)

$$\Pi = 3\,\frac{O}{a^3} = 29^m{,}345 \cdot \frac{1}{a}$$

where a denotes the radius of the terrestrial equator to be expressed in metres; therefore

$$\log a = 6{,}80366$$

which gives for the factor in question

$$1 - \frac{\lambda}{4254200}$$

a very small correction but of greater effect on longer than on shorter pendulums, and always diminishing the time of oscillation.

The term in the differential of κ, which we have just considered, is the greatest, the others arising from the ellipticity and centrifugal force of the earth; but as they give occasion to other forms, it will be useful to investigate them.

Let us first take the two parts

$$\left(\frac{d\kappa}{d\psi}\right)=-2\Lambda\theta\sin\psi;\ \theta\left(\frac{d\kappa}{d\theta}\right)=2\Lambda\theta\cos\psi.$$

Before we substitute these values in the differential equations of $\varepsilon, \varepsilon', \eta$ and α, we must find the expressions for $\theta\sin\psi$ and $\theta\cos\psi$.

From the equation (F_2) we have, after introducing η,

$$\tan(\psi-M)=\frac{\varepsilon'}{\varepsilon}\tan(\omega t+\eta),$$

therefore

$$\theta\sin(\psi-M)=\varepsilon'\sin(\omega t+\eta)$$
$$\theta\cos(\psi-M)=\varepsilon\cos(\omega t+\eta)$$

where as before,

$$M=\alpha+(\mu+\tfrac{1}{2}\mu')\,t$$

whence

$$\theta\sin\psi=\varepsilon\sin M\cos(\omega t+\eta)+\varepsilon'\cos M\sin(\omega t+\eta)$$
$$\theta\cos\psi=\varepsilon\cos M\cos(\omega t+\eta)-\varepsilon'\sin M\sin(\omega t+\eta).$$

With regard to these values, the substitution of the partial values of $\left(\frac{d\kappa}{d\psi}\right)$ and $\theta\left(\frac{d\kappa}{d\theta}\right)$ give the following differential equations:

$$\frac{d\varepsilon}{dt}=-\frac{\Lambda}{\omega}\cos M\sin(\omega t+\eta)$$

$$\frac{d\varepsilon'}{dt}=-\frac{\Lambda}{\omega}\sin M\cos(\omega t+\eta)$$

$$\frac{d\eta}{dt}=-\frac{\Lambda\varepsilon}{\omega(\varepsilon^2-\varepsilon'^2)}\cos M\cos(\omega t+\eta)-\frac{\Lambda\varepsilon'}{\omega(\varepsilon^2-\varepsilon'^2)}\sin M\sin(\omega t+\eta)$$

$$\frac{d\alpha}{dt}=\frac{\Lambda\varepsilon}{\omega(\varepsilon^2-\varepsilon'^2)}\sin M\sin(\omega t+\eta)+\frac{\Lambda\varepsilon'}{\omega(\varepsilon^2-\varepsilon'^2)}\cos M\cos(\omega t+\eta)$$

whence we derive, neglecting the small factor $(\mu+\frac{1}{2}\mu')$, and taking on the right hand side $\eta=0$, the following integrals of the variations:

$$\delta\varepsilon=\frac{\Lambda}{\omega^2}\cos M\cos\omega t$$

$$\delta\varepsilon'=-\frac{\Lambda}{\omega^2}\sin M\sin\omega t$$

$$\delta\eta=-\frac{\Lambda\varepsilon}{\omega^2(\varepsilon^2-\varepsilon'^2)}\cos M\sin\omega t+\frac{\Lambda\varepsilon'}{\omega^2(\varepsilon^2-\varepsilon'^2)}\sin M\cos\omega t$$

$$\delta\alpha=-\frac{\Lambda\varepsilon}{\omega^2(\varepsilon^2\pm\varepsilon'^2)}\sin M\cos\omega t+\frac{\Lambda\varepsilon'}{\omega^2(\varepsilon^2-\varepsilon'^2)}\cos M\sin\omega t$$

These terms are extremely small, they exert, however, on the motion of the pendulum a very remarkable though imperceptible effect, which consists in soliciting the pendulum to perform its oscillations, not about the plumb line, but about an other, deviating from it towards north. In order to show this, we will introduce again the coordinates

$$x=\theta\cos\psi\,;\ y=\theta\sin\psi.$$

By means of the preceding values of $\theta\cos\psi$ and $\theta\sin\psi$, we obtain

$$\begin{aligned}\delta x=&\delta\varepsilon\cos M\cos\omega t-\delta\varepsilon'\sin M\sin\omega t\\&-\delta\eta\,(\varepsilon\cos M\sin\omega t+\varepsilon'\sin M\cos\omega t)\\&-\delta\alpha\,(\varepsilon\sin M\cos\omega t+\varepsilon'\cos M\sin\omega t)\\\delta y=&\delta\varepsilon\sin M\cos\omega t+\delta\varepsilon'\cos M\sin\omega t\\&-\delta\eta\,(\varepsilon\sin M\sin\omega t-\varepsilon'\cos M\cos\omega t)\\&+\delta\alpha\,(\varepsilon\cos M\cos\omega t-\varepsilon'\sin M\sin\omega t)\end{aligned}$$

and substituting the values of $\delta\varepsilon$, $\delta\varepsilon'$, $\delta\eta$ and $\delta\alpha$, we obtain

$$\delta x=\frac{\Lambda}{\omega^2}\,;\ \delta y=0$$

whence we see that all the abscissas are changed by the quantity $\frac{\Lambda}{\omega^2}$, or, since Λ is negative, are shifted towards north, while the ordinates undergo no variation; so that if the origin of these co-ordinates is transferred towards north by the quantity $-\frac{\Lambda}{\omega^2}$, the motion of the pendulum about the new origin becomes symmetrical. The trigonometrical tangent of the angle which the plumbline makes with the straight line passing through this new origin, and the point of suspension, is expressed by the quantity $-\frac{\Lambda}{\omega^2}$. Taking as before

$$\omega^2 = \frac{\lambda}{g},$$

and substituting the value of Λ (page 172), we have

$$-\frac{\Lambda}{\omega^2} = \frac{4\,n^2 - 3\frac{\mathrm{P}}{a^5}}{g}\,\lambda \sin\phi \cos\phi$$

which by means of the numerical values (page 169) gives for the maximum of the corresponding angle the very small quantity

$$0'',0002\,\lambda.$$

In order to substitute the partial values

$$\left(\frac{d\,\kappa}{d\,\psi}\right) = 2\,\Gamma\theta \sin\psi\,.\,\theta\cos\psi;\ \theta\left(\frac{d\,\kappa}{d\,\theta}\right) = -2\,\Gamma\theta\cos\psi\,.\,\theta\cos\psi$$

we multiply the second members of the last differential equations of ε, ε', η and α, by the expression for $\theta \cos\psi$, viz. —

$$\theta\cos\psi = \varepsilon\cos\mathrm{M}\,\cos(\omega t + \eta) - \varepsilon' \sin\mathrm{M}\sin(\omega t + \eta)$$

and write $-\Gamma$ instead of Λ. Thus we obtain, making at the same time $\eta=0$, the following differential equations:—

$$\frac{d\epsilon}{dt}=-\frac{\Gamma}{4\omega}\epsilon'\sin 2M+\frac{\Gamma}{2\omega}\epsilon\cos^2 M\sin 2\omega t+\frac{\Gamma}{2\omega}\epsilon'\sin M\cos M\cos 2\omega t$$

$$\frac{d\epsilon'}{dt}=\frac{\Gamma}{4\omega}\epsilon\sin 2M-\frac{\Gamma}{2\omega}\epsilon'\sin^2 M\sin 2\omega t+\frac{\Gamma}{2\omega}\epsilon\sin M\cos M\cos 2\omega t$$

$$\frac{d\eta}{dt}=\frac{\Gamma}{4\omega}+\frac{\Gamma(\epsilon^2+\epsilon'^2)}{4\omega(\epsilon^2-\epsilon'^2)}\cos 2M+\frac{\Gamma}{2\omega(\epsilon^2-\epsilon'^2)}(\epsilon^2\cos^2 M+\epsilon'^2\sin^2 M)\cos 2\omega t$$

$$\frac{d\alpha}{dt}=-\frac{\Gamma\epsilon\epsilon'}{2\omega(\epsilon^2-\epsilon'^2)}\cos 2M-\frac{\Gamma\epsilon\epsilon'}{2\omega(\epsilon^2-\epsilon'^2)}\cos 2\omega t-\frac{\Gamma}{2\omega}\sin M\cos M\sin 2\omega t$$

Since the terms affected with the factor $\sin 2\omega t$, or $\cos 2\omega t$, can never become considerable, they will be omitted in integrating, likewise the indeterminate quantity μ' will be neglected, so that

$$M=\alpha+\mu t=\alpha+n t\sin\phi$$

and the arbitrary constants to be added to the integrals of the preceding equations will be so determined that the increments $\delta\epsilon$, $\delta\epsilon'$, $\delta\eta$ and $\delta\alpha$ vanish for $t=0$. Thus we obtain

$$\left.\begin{aligned}
\delta\epsilon&=-\frac{\Gamma\epsilon'}{8\omega n\sin\phi}[\cos 2\alpha-\cos 2(\alpha+\mu t)]\\
\delta\epsilon'&=\frac{\Gamma\epsilon}{8\omega n\sin\phi}[\cos 2\alpha-\cos 2(\alpha+\mu t)]\\
\delta\eta&=\frac{\Gamma}{4\omega}t+\frac{\Gamma(\epsilon^2+\epsilon'^2)}{8\omega n\sin\phi(\epsilon^2-\epsilon'^2)}[\sin 2\alpha-\sin 2(\alpha+\mu t)]\\
\delta\alpha&=\frac{\Gamma\epsilon\epsilon'}{4\omega n\sin\phi(\epsilon^2-\epsilon'^2)}[\sin 2\alpha-\sin 2(\alpha+\mu t)].
\end{aligned}\right\}(1)$$

Here we have terms which depend on the azimuth in which the pendulum commences to swing, and which therefore show that the primitive azimuth of the plane of oscillation exerts an influence upon the motion of the latter.

Since the quantity $\sin\phi$ occurs in the denominators of these expressions, the latter will be of no use if the place of observation is at the equator. In this case $\mu = 0$, and the integrals therefore assume another form.

With regard to places near the equator, where μt remains a small arc, even in experiments of long duration, we are able to expand $\sin 2(\alpha + \mu t)$ and $\cos(\alpha + \mu t)$ into highly converging series, by means of which the integrals (1) are transformed into the following:

$$\left.\begin{aligned}
\delta \epsilon &= -t \frac{\Gamma}{4\omega} \epsilon' \sin 2\alpha - t^2 \frac{\Gamma n \sin\phi}{4\omega} \epsilon' \cos 2\alpha + \&c. \\
\delta \epsilon' &= t \frac{\Gamma}{4\omega} \epsilon \sin 2\alpha + t^2 \frac{\Gamma n \sin\phi}{4\omega} \epsilon \cos 2\alpha + \&c. \\
\delta \eta &= t \frac{\Gamma}{4\omega} + t \frac{\Gamma(\epsilon^2 + \epsilon'^2)}{4\omega(\epsilon^2 - \epsilon'^2)} \cos 2\alpha - t^2 \frac{\Gamma n \sin\phi (\epsilon^2 + \epsilon'^2)}{4\omega(\epsilon^2 - \epsilon'^2)} \sin 2\alpha + \&c. \\
\delta \alpha &= -t \frac{\Gamma \epsilon \epsilon'}{2\omega(\epsilon^2 - \epsilon'^2)} \cos 2\alpha + t^2 \frac{\Gamma n \sin\phi\, \epsilon \epsilon'}{2\omega(\epsilon^2 - \epsilon'^2)} \sin 2\alpha + \&c.
\end{aligned}\right\} (2.)$$

All the terms except those multiplied by the first power of t vanish if $\phi = 0$, and the latter therefore express the values of $\delta \epsilon$, $\delta \epsilon'$, $\delta \eta$ and $\delta \alpha$, when the point of suspension of the pendulum is over the equator.

It is not necessary to discuss any further the integrals just found; they are of no importance, since there exist other terms much greater, and yet scarcely exerting a perceptible influence. We will now proceed to determine these greater terms.

In the several preceding integrations we have considered the quantities ϵ, ϵ', η, and α as constants on the right-hand side of the equations, which is not strictly true, since they, in consequence of the method for the variation of arbitrary constants, have become variable. But on the assumption of these quantities being invariable, only very small terms have been added to their values, resulting from the first approximation, and their variation therefore may be found by a method similar to that by which we determine the square of the disturbing force in the theory of planetary perturbations.

The greatest terms which we can obtain in this case result from a combination of the integrals (1) with the first terms of the values of the differentials of η and α, viz.:

$$\frac{d\eta}{dt} = -\tfrac{1}{16}\,\omega\,(\varepsilon^2 + \varepsilon'^2)$$

$$\frac{d\alpha}{dt} = \tfrac{1}{24}\,\omega\,\varepsilon\varepsilon'$$

Considering here ε and ε' as variable quantities which take the increments $\delta\varepsilon$ and $\delta\varepsilon'$, we have

$$\frac{d\eta}{dt} = -\tfrac{1}{8}\,\omega\,(\varepsilon\,\delta\varepsilon + \varepsilon'\,\delta\varepsilon')$$

$$\frac{d\alpha}{dt} = \tfrac{1}{24}\,\omega\,(\varepsilon'\,\delta\varepsilon + \varepsilon\,\delta\varepsilon').$$

Substituting in these expressions the values of $\delta\varepsilon$ and $\delta\varepsilon'$ from the first two equations of (1), we obtain

$$\frac{d\eta}{dt} = 0$$

$$\frac{d\alpha}{dt} = \frac{\Gamma}{192\,n\sin\phi}\,(\varepsilon^2 - \varepsilon'^2)\,[\cos 2\alpha - \cos 2(\alpha + \mu t)].$$

By integration therefore determining the arbitrary constants to be added so that the integrals vanish for $t = 0$, we have

$$\delta\eta = 0$$

$$\delta\alpha = \frac{t\,\Gamma\,(\varepsilon^2 - \varepsilon'^2)}{192\,n\sin\phi}\cos 2\alpha + \frac{\Gamma\,(\varepsilon^2 - \varepsilon'^2)}{384\,n^2\sin^2\phi}\,[\sin 2\alpha - \sin 2(\alpha + \mu t)].$$

We see that our more exact determination exerts no influence upon the value of η; but the effect on the value of α is far greater than that indicated by the equations (1) and (2), for we have now in the denominator of the second term the square of the small quantity n, and moreover the expression is freed from the general factor ε'.

If the place of observation is near the equator, we have, by expanding $\sin 2(\alpha+\mu t)$ in an infinite series,

$$\delta\alpha = t^2 \frac{\Gamma}{192}(\epsilon^2-\epsilon'^2)\sin 2\alpha + t^3 \frac{\Gamma n \sin\phi}{288}(\epsilon^2-\epsilon'^2)\cos 2\alpha + \ \&c.$$

therefore at the equator

$$\delta\alpha = t^2 \frac{\Gamma}{192}(\varepsilon^2-\varepsilon'^2)\sin 2\alpha.$$

In order to illustrate the magnitude of the effect of this term, we will take a numerical example.

Let $\varepsilon=\frac{1}{9}$ and ε' inappreciable, then by means of the value of Γ we have

$$\delta\alpha = t^2\,(3{,}1383-10)\sin 2\alpha$$

where the number within the brackets denotes the logarithm of the coefficient expressed in seconds. Supposing the pendulum-experiment lasted as long as 6 hours, then

$$t=6\times 3600$$

whence

$$\delta\alpha = 64'' \sin 2\alpha.$$

Hence the greatest change of the motion of the plane of oscillation at the equator generated by the terms just developed during the time of 6 hours amounts to only

64″, and takes place when the initial azimuth of the plane of oscillation is either 45° or 225°, or 135° or 315°. In the first two cases this change is positive, in the other two it is negative. This change will be zero if the pendulum commences to swing either in the meridian or in the direction perpendicular to it.

For shortness, let

$$R=\frac{t\,\Gamma}{192\,n\sin\phi}(\varepsilon^2-\varepsilon'^2)\,; \qquad S=\frac{\Gamma}{384\,n^2\sin^2\phi}(\varepsilon^2-\varepsilon'^2)$$

then the maximum of $\delta\alpha$ takes place for any latitude ϕ and duration t, if

$$\tan 2\,\alpha=\frac{S\,(1-\cos 2\,\mu\,t)}{R-S\sin 2\,\mu\,t}$$

and the maximum itself is therefore expressed by

$$\sqrt{R^2-R\,S\sin 2\,\mu\,t+4\,S^2\sin^2\mu\,t}$$

whence we see that if t is not greater than 6 h., the maximum maximorum takes place for $\phi=0$, id est, at the equator, so that for any other latitude

$$\delta\,\alpha \angle 64''.$$

Considering the length of duration of the experiment we have here assumed, this deviation is so small that perhaps the most accurate observation will be scarcely able to ascertain it.

We now substitute the remaining third term of the value of $\theta\left(\frac{d\,\kappa}{d\,\theta}\right)$ into the differential equation (a) and integrate; thus we obtain

$$\delta\,\epsilon = -\frac{\Lambda\,\epsilon^2}{6\,\omega^2}\cos M\cos 3\,\omega\,t - \frac{\Lambda\,\epsilon^2}{2\,\omega^2}\cos M\cos\omega\,t$$

$$+\frac{\Lambda\,\epsilon\,\epsilon'}{6\,\omega^2}\sin M\sin 3\,\omega\,t - \frac{\Lambda\,\epsilon\,\epsilon'}{2\,\omega^2}\sin M\sin\omega\,t$$

$$\delta\,\epsilon' = \frac{\Lambda\,\epsilon\,\epsilon'}{6\,\omega^2}\cos M\cos 3\,\omega\,t + \frac{\Lambda\,\epsilon\,\epsilon'}{2\,\omega^2}\cos M\cos\omega\,t$$

$$-\frac{\Lambda\,\epsilon'^2}{6\,\omega^2}\sin M\sin 3\,\omega\,t + \frac{\Lambda\,\epsilon'^2}{2\,\omega^2}\sin M\sin\omega\,t$$

$$\delta\,\eta = \frac{\Lambda\,\epsilon}{\omega^2}\cos M\sin\omega\,t + \frac{\Lambda\,\epsilon'}{\omega^2}\sin M\cos\omega\,t + \frac{\Lambda\,\epsilon\,(\epsilon^2+\epsilon'^2)}{6\,\omega^2\,(\epsilon^2-\epsilon'^2)}\cos M\sin 3\,\omega\,t$$

$$+\frac{\Lambda\,\epsilon\,(\epsilon^2+\epsilon'^2)}{2\,\omega^2\,(\epsilon^2-\epsilon'^2)}\cos M\sin\omega\,t + \frac{\Lambda\,\epsilon'\,(\epsilon^2+\epsilon'^2)}{6\,\omega^2\,(\epsilon^2-\epsilon'^2)}\sin M\cos 3\,\omega\,t$$

$$-\frac{\Lambda\,\epsilon'\,(\epsilon^2+\epsilon'^2)}{2\,\omega^2\,(\epsilon^2-\epsilon'^2)}\sin M\cos\omega\,t$$

$$\delta\,a = -\frac{\Lambda\,\epsilon^2\,\epsilon'}{6\,\omega^2\,(\epsilon^2-\epsilon'^2)}\cos M\sin 3\,\omega\,t - \frac{\Lambda\,\epsilon^2\,\epsilon'}{2\,\omega^2\,(\epsilon^2-\epsilon'^2)}\cos M\sin\omega\,t$$

$$-\frac{\Lambda\,\epsilon\,\epsilon'^2}{6\,\omega^2\,(\epsilon^2-\epsilon'^2)}\sin M\cos 3\,\omega\,t + \frac{\Lambda\,\epsilon\,\epsilon'^2}{2\,\omega^2\,(\epsilon^2-\epsilon'^2)}\sin M\cos\omega\,t.$$

The influence of these expressions is imperceptible.

It remains, finally, to consider the second lines of the differential equations (a), but the terms arising from it are likewise extremely small, and it may therefore suffice to state the results.

By means of the equations

$$\theta^2 = \varepsilon^2\cos^2\omega\,t + \varepsilon'^2\sin^2\omega\,t$$

$$\theta\frac{d\,\theta}{dt} = -\omega(\varepsilon^2-\varepsilon'^2)\sin\omega\,t\cos\omega\,t$$

$$\theta^2\frac{d\,\psi}{dt} = \omega\,\varepsilon\,\varepsilon' + \mu\,\theta^2$$

where μ' is neglected in the last equation, we find by substitution and reduction

$$\theta\frac{d\,\psi}{d\,t}\,\varepsilon\cos\omega\,t + \frac{d\,\theta}{d\,t}\,\varepsilon'\sin\omega\,t = \theta\,\omega\,\varepsilon'\cos\omega\,t + \theta\,\mu\,\varepsilon\cos\omega t$$

$$\theta\frac{d\,\psi}{d\,t}\,\varepsilon'\sin\omega\,t - \frac{d\,\theta}{d\,t}\,\varepsilon\cos\omega\,t = \theta\,\omega\,\varepsilon\sin\omega\,t + \theta\,\mu\varepsilon'\sin\omega\,t$$

$$\theta \frac{d\psi}{dt}(\varepsilon^2 \cos^2 \omega t - \varepsilon'^2 \sin^2 \omega t) + 2 \frac{d\theta}{dt} \varepsilon \varepsilon' \sin \omega t \cos \omega t$$
$$= \theta \omega (\varepsilon \varepsilon' - 2 \varepsilon \varepsilon' \sin^2 \omega t) + \theta \mu (\varepsilon^2 \cos^2 \omega t - \varepsilon'^2 \sin^2 \omega t)$$
$$\theta \frac{d\psi}{dt} \varepsilon \varepsilon' (\cos^2 \omega t - \sin^2 \omega t) + \frac{d\theta}{dt}(\varepsilon^2 + \varepsilon'^2) \sin \omega t \cos \omega t$$
$$= -\theta \omega (\varepsilon^2 \sin^2 \omega t - \varepsilon'^2 \cos^2 \omega t) + \theta \mu \varepsilon \varepsilon' \cos 2 \omega t$$

whence by integration and neglecting the factor μ, as small compared to ω, we have

$$\delta \epsilon = -\frac{2n \cos \phi}{3\omega}[\epsilon \epsilon' \cos M \cos^3 \omega t + \epsilon'^2 \sin M \sin^3 \omega t]$$

$$\delta \epsilon' = \frac{2n \cos \phi}{3\omega}[\epsilon^2 \cos M \cos^3 \omega t + \epsilon \epsilon' \sin M \sin^3 \omega t]$$

$$\delta \eta = \frac{n \cos \phi}{\omega(\epsilon^2 - \epsilon'^2)}\left[\begin{array}{l}\epsilon^2 \epsilon' \cos M \sin \omega t - \epsilon \epsilon'^2 \sin M \cos \omega t \\ + \frac{1}{3} \epsilon^2 \epsilon' \cos M \sin 3 \omega t + \frac{1}{3} \epsilon \epsilon'^2 \sin M \cos 3 \omega t\end{array}\right]$$

$$\delta a = \frac{n \cos \phi}{\omega(\epsilon^2 - \epsilon'^2)}\left[\begin{array}{l}\frac{1}{2} \epsilon (\epsilon^2 - 3\epsilon'^2) \cos M \sin \omega t + \frac{1}{2} \epsilon' (3\epsilon^2 - \epsilon'^2) \sin M \cos \omega t \\ -\frac{1}{6} \epsilon (\epsilon^2 + \epsilon'^2) \cos M \sin 3 \omega t - \frac{1}{6} \epsilon' (\epsilon^2 + \epsilon'^2) \sin M \cos 3 \omega t\end{array}\right]$$

Summing up the results of all these developements, we observe that of all the expressions found in the second approximation, only one term is of appreciable value, viz. —

$$\delta \alpha = \tfrac{1}{24} \omega \varepsilon \varepsilon' t.$$

We have shown by a numerical example that the latter may become considerable under circumstances which can not always be avoided.

Its effect, however, is modified by the circumstance that the resistance of the air is continually shortening the arcs of the greatest oscillations of the pendulum, and it is therefore necessary to consider the influence of this resistance.

Effect of the resistance of the air on the motion of the free pendulum.

The resistance which a medium opposes to a body

moving in it, is a force acting in the direction of the tangent to the curve which that body describes, and vanishing if the relative velocity of the body and the medium is zero; it is therefore a function of the relative velocity. Moreover this force is also a function of the figure of the body, its density and the density of the medium.

Taking here into account only the resistance of the atmosphere, which moves uniformly with the earth, we may assume

$$d s^2 = d \xi^2 + d \eta^2 + d \zeta^2$$

where $\frac{d s}{d t}$ expresses the velocity of which the resistance of the air is a function. Denoting by W a function of $\frac{d s}{d t}$, of the figure of the body, of its density and of that of the air, we have for the components of the resisting force acting parallel to the axes ξ, η, ζ the expressions

$$-\mathrm{W}\frac{d \xi}{d t}; \; -\mathrm{W}\frac{d \eta}{d t}; \; -\mathrm{W}\frac{d \zeta}{d t}$$

and these are the quantities which must be added to the second members of the equations (C) if the effect of the resisting medium is taken into consideration.

For the same reason we must add the quantities

$$\mathrm{W}\left(\eta \frac{d \xi}{d t} - \xi \frac{d \eta}{d t}\right); \; \mathrm{W}\left(\xi \frac{d \zeta}{d t} - \zeta \frac{d \xi}{d t}\right); \; \mathrm{W}\left(\zeta \frac{d \eta}{d t} - \eta \frac{d \zeta}{d t}\right)$$

to the right-hand sides of the equations (D) and (E); but the equation (F) would by this means become an incomplete integral.

In order, therefore, to find the additions to be made to

the integrals (A_1) (B_1) and (C_1) we take the equation (L), viz—

$$C\,dr\,dt+(B-A)\,pq\,dt^2=\int[\gamma\,(\eta\,d^2\zeta-\zeta\,d^2\eta)+\gamma'\,(\zeta\,d^2\xi-\xi\,d^2\zeta)$$
$$+\gamma''\,(\xi\,d^2\eta-\eta\,d^2\xi)]\,dm$$

and in a similar way as we have derived this equation, or by merely interchanging letters, we obtain the following two —

$$B\,dq\,dt+(A-C)\,pr\,dt^2=\int[\beta\,(\eta\,d^2\zeta-\zeta\,d^2\eta)+\beta'\,(\zeta\,d^2\xi-\xi\,d^2\zeta)$$
$$+\beta''\,(\xi\,d^2\eta-\eta\,d^2\xi)]\,dm$$

$$A\,dp\,dt+(C-A)\,qr\,dt^2=\int[\alpha\,(\eta\,d^2\zeta-\zeta\,d^2\eta)+\alpha'\,(\zeta\,d^2\xi-\xi\,d^2\zeta)$$
$$+\alpha''\,(\xi\,d^2\eta-\eta\,d^2\xi)]\,dm.$$

Considering only those terms which arise from the resistance of the air, and making at the same time A = B, the preceding three equations become

$$C\,dr\,.\;.\;.\;.\;.\;.\;.=\int W\,[(\gamma''\eta-\gamma'\zeta)\,d\xi+(\gamma\zeta-\gamma''\xi)\,d\eta$$
$$+(\gamma'\xi-\gamma\eta)\,d\xi]\,dm$$

$$A\,dq+(A-C)\,pr\,dt=\int W\,[(\beta''\eta-\beta'\zeta)\,d\xi+(\beta\zeta-\beta''\xi)\,d\eta$$
$$+(\beta'\xi-\beta\eta)\,d\zeta]\,dm$$

$$A\,dp-(A-C)\,qr\,dt=\int W\,[(\alpha''\eta-\alpha'\zeta)\,d\xi+(\alpha\zeta-\alpha''\xi)\,d\eta$$
$$+(\alpha'\xi-\alpha\eta)\,d\zeta]\,dm.$$

From the equations (G) we easily obtain

$$\gamma''\eta-\gamma'\zeta=\alpha\,v-\beta\,u;\;\gamma\zeta-\gamma''\xi=\alpha'v-\beta'u;$$
$$\gamma'\xi-\gamma\,\eta=\alpha''v-\beta''u;$$
$$\beta''\eta-\beta'\zeta=\gamma\,u-\alpha\,w;\;\beta\,\zeta-\beta''\xi=\gamma'u-\alpha'w;$$
$$\beta'\xi-\beta\,\eta=\gamma''u-\alpha''w;$$
$$\alpha''\eta-\alpha'\zeta=\beta\,w-\gamma\,v;\;\alpha\,\zeta-\alpha''\zeta=\beta'w-\gamma'v;$$
$$\alpha'\xi-\alpha\,\eta=\beta''w-\gamma''v.$$

Hence by means of the equations (H)

$$C\frac{dr}{dt} \quad . \quad . \quad . \quad . = -r\int W(u^2+v^2)\,dm + p\int W\,u\,w\,dm + q\int W\,v\,w\,dm$$

$$A\frac{dq}{dt} + (A-C)\,p\,r = -q\int W(u^2+w^2)\,dm + r\int W\,v\,w\,dm + p\int W\,u\,v\,dm$$

$$A\frac{dp}{dt} - (A-C)\,q\,r = -p\int W(v^2+w^2)\,dm + q\int W\,u\,v\,dm + r\int W\,u\,w\,dm.$$

Since a well constructed pendulum must be symmetrical about the axis of w, we have on this supposition

$$\int W\,u\,v\,dm=0;\ \int W\,v\,w\,dm=0;\ \int W\,u\,w\,dm=0$$

$$\int W(u^2+w^2)\,dm=\int W(v^2+w^2)\,dm.$$

Assuming

$$\int W(u^2+v^2)\,dm=C\,k';\ \int W(u^2+w^2)\,dm=A\,k$$

we have by substitution

$$\frac{dr}{dt} \quad . \quad . \quad . \quad = -k'r$$

$$\frac{dq}{dt}+\frac{A-C}{A}\,p\,r=-k\,q$$

$$\frac{dp}{dt}-\frac{A-C}{A}\,q\,r=-k\,p.$$

We may here remark that though the effect of the resistance on the angular velocity r, can under these conditions not become zero, yet k' is much smaller than k, the more so the smoother the surface of the pendulum.

Let us first take the hypothesis that the resistance of the air is proportional to the first power of the velocity, then k' and k are both constant.

The integral of the equation $d\,r=-k'r\,d\,t$ is therefore

$$r=n'e^{-k't}$$

where e denotes the base of the hyperbolic logarithms and n' is the arbitrary constant having the same meaning as in the integral ($A_{,}$).

The equation shows that the resistance of the air counteracts the rotation of the pendulum about its axis; but since k' is a very small quantity, the exponential function can only exert a perceptible effect after the lapse of a long time, and we may therefore neglect it for the short duration of a pendulum experiment. Consequently, we take $r = n'$ and the expression $(A_{,})$ remains unaltered.

But the case is different with the other two differential equations, viz—

$$\frac{dq}{dt} + \frac{A - C}{A} p r = -k q$$

$$\frac{dp}{dt} - \frac{A - C}{A} q r = -k p$$

because k is much greater than k'. Multiplying the first of these two equations by $q\,dt$ the other by $p\,dt$, then adding and integrating, we have

$$p^2 + q^2 = -2 \int k (p^2 + q^2)\, dt$$

and we must therefore add to the right-hand side of the equation $(B_{,})$ the quantity

$$-2 A \int k (p^2 + q^2)\, dt.$$

Multiplying the last three differential equations respectively by $C \gamma'' dt$, $A \beta'' dt$, and $A \alpha'' dt$, then adding and integrating, we obtain, by means of the equations derived from (G);

$$A (\alpha'' dp + \beta'' dq - r\, d\gamma'') + C (\gamma'' dr + r\, d\gamma'')$$
$$= -A k (\alpha'' p + \beta'' q)\, dt - C k' \gamma'' r\, dt$$

but the same equations derived from (G) give

$$A (p\, d\alpha'' + q\, d\beta'' + r\, d\gamma'') = 0$$

therefore adding this equation to the preceding, and integrating,

$$A (\alpha'' p + \beta'' q) + C \gamma'' r = -A \int k (\alpha'' p + \beta'' q)\, dt - C \int k' \gamma'' r\, dt.$$

Neglecting k' as small compared to k, it follows that the quantity to be added to the right-hand side of the integral ($C_{,}$) is

$$-A \int k\,(\alpha'' p+\beta'' q)\, dt.$$

Consequently, the quantities κ and κ' determined by the equations ($G_{,}$) must on account of the resistance of the air receive the following terms:

$$\kappa = -2 \int k\,(p^2+q^2)\, dt$$
$$\kappa' + - \int k\,(\alpha'' p+\beta'' q)\, dt$$

and for the same reason the values of D and E in our second approximation receive the additions

$$D = -k \frac{\theta}{d\theta}\left[p^2+q^2-\frac{d\psi}{dt}(\alpha'' p+\beta'' q)\right]$$
$$E = -k\,(\alpha'' p+\beta'' q).$$

Eliminating the known values of p, q, &c., and neglecting the higher powers of θ, we obtain

$$D = -k\theta \frac{d\theta}{dt} = k\omega\,(\varepsilon^2-\varepsilon'^2)\sin(\omega t+\eta)\cos(\omega t+\eta)$$
$$E = -k\theta^2 \frac{d\psi}{dt} = -k\,(\mu+\tfrac{1}{2}\mu')\,\theta^2 - k\omega\varepsilon\varepsilon'$$

and substituting these values into the first expressions for the differentials of ε, ε', η and α, we have

$$\left.\begin{aligned}
\frac{d\varepsilon}{dt} &= -k\left[\varepsilon+\frac{\varepsilon'}{\omega}(\mu+\tfrac{1}{2}\mu')\right]\sin^2(\omega t+\eta)\\
\frac{d\varepsilon'}{dt} &= -k\left[\varepsilon'+\frac{\varepsilon}{\omega}(\mu+\tfrac{1}{2}\mu')\right]\cos^2(\omega t+\eta)\\
\frac{d\eta}{dt} &= -k\,\frac{(\varepsilon^2+\varepsilon'^2)}{(\varepsilon^2-\varepsilon'^2)}\sin(\omega t+\eta)\cos(\omega t+\eta)\\
&\quad -\frac{2k\varepsilon\varepsilon'(\mu+\tfrac{1}{2}\mu')}{\omega(\varepsilon^2-\varepsilon'^2)}\sin(\omega t+\eta)\cos(\omega t+\eta)\\
\frac{d\alpha}{dt} &= 2k\,\frac{\varepsilon\varepsilon'}{(\varepsilon^2-\varepsilon'^2)}\sin(\omega t+\eta)\cos(\omega t+\eta)\\
&\quad +\frac{k(\varepsilon^2+\varepsilon'^2)(\mu+\tfrac{1}{2}\mu')}{\omega(\varepsilon^2-\varepsilon'^2)}\sin(\omega t+\eta)\cos(\omega t+\eta)
\end{aligned}\right\}\quad (b.)$$

But

$$\varepsilon' = \frac{\nu - \mu - \frac{1}{2}\mu'}{\omega} \varepsilon.$$

Substituting this value in the second of the preceding equations, we have

$$\frac{d\varepsilon'}{dt} = -\nu \frac{k\varepsilon}{\omega} \cos^2 (\omega t + \eta)$$

whence the following theorem.

Whatever be the law of the resistance of the air, the least arcs of oscillations of the free pendulum undergo no change in consequence of this resistance, unless the pendulum received a lateral impulse at the commencement of its motion.

If we take each of the quantities ε' μ, μ' equal to zero, the four preceding differential equations become integrable. The first and third of them are transformed into the following:

$$\frac{d\varepsilon}{dt} = -k\varepsilon \sin^2 (\omega t + \eta)$$

$$\frac{d\eta}{dt} = -k \sin (\omega t + \eta) \cos (\omega t + \eta).$$

In order to integrate them we assume

$$p = \varepsilon \sin \eta; \quad q = \varepsilon \cos \eta$$

therefore

$$\frac{dp}{dt} = \frac{d\varepsilon}{dt} \sin \eta + \frac{d\eta}{dt} \varepsilon \cos \eta; \quad \frac{dq}{dt} = \frac{d\varepsilon}{dt} \cos \eta - \frac{d\eta}{dt} \varepsilon \sin \eta$$

or substituting the preceding values of $d\varepsilon$ and $d\eta$

$$\frac{dp}{dt} = -k\varepsilon \sin (\omega t + \eta) \cos \omega t; \quad \frac{dq}{dt} = -k\varepsilon \sin (\omega t + \eta) \sin \omega t$$

and eliminating $\varepsilon \sin \eta$ and $\varepsilon \cos \eta$, we obtain

$$\frac{dp}{dt} = -\tfrac{1}{2} k p - \tfrac{1}{2} k p \cos 2 \omega t - \tfrac{1}{2} k q \sin 2 \omega t$$

$$\frac{dq}{dt} = -\tfrac{1}{2} k q + \tfrac{1}{2} k q \cos 2 \omega t - \tfrac{1}{2} k p \sin 2 \omega t.$$

In order to simplify the integration of these equations we must transform them into others in which the coefficients of the variables are constant quantities. This is effected by introducing two new variables C and D, which are connected with p and q, by the following equations: —

$$\left.\begin{aligned} C &= -e^{\frac{1}{2}kt} p \sin \omega t + e^{\frac{1}{2}kt} q \cos \omega t \\ D &= e^{\frac{1}{2}kt} p \cos \omega t + e^{\frac{1}{2}kt} q \sin \omega t \end{aligned}\right\} \quad . \quad . \quad . \quad (\alpha)$$

where e as before denotes the base of the hyperbolic logarithms. Differentiating these equations and substituting the preceding values of dp and dq, then we obtain

$$\left.\begin{aligned} \frac{dC}{dt} &= \tfrac{1}{2} k C - \omega D \\ \frac{dD}{dt} &= -\tfrac{1}{2} k D + \omega C \end{aligned}\right\} \quad . \quad . \quad . \quad . \quad . \quad . \quad . \quad . \quad (\beta)$$

The differentiation of the first of these equations gives

$$\frac{d^2C}{dt^2} = \tfrac{1}{2} k \frac{dC}{dt} - \omega \frac{dD}{dt},$$

but multiplying the two equations (β) by $\tfrac{1}{2} k$ and ω respectively, we get by subtraction

$$\tfrac{1}{2} k \frac{dC}{dt} - \omega \frac{dD}{dt} = (\tfrac{1}{4}k^2 - \omega^2) C$$

Therefore

$$\frac{d^2C}{d\,t^2} = -(\omega^2 - \tfrac{1}{4}k^2)\,C.$$

Let us assume, since $\omega > k$, that

$$\Omega^2 = \omega^2 - \tfrac{1}{4}k^2$$

then we have by integration

$$C = \lambda \cos \Omega\,t + \lambda' \sin \Omega\,t \quad . \quad . \quad . \quad . \quad (\gamma)$$

where λ and λ' denote the two arbitrary constants.

Substituting the first differential of this integral, viz.—

$$\frac{dC}{dt} = -\Omega\lambda \sin \Omega\,t + \Omega\lambda' \cos \Omega t$$

and also the integral itself into the first equation of (β), then we obtain the integral of the second equation of (β)

$$D = \left(\frac{k}{2\,\omega}\lambda + \frac{\Omega}{\omega}\lambda'\right) \cos \Omega\,t + \left(\frac{k}{2\,\omega}\lambda' - \frac{\Omega}{\omega}\lambda\right) \sin \Omega\,t\,. \;(\delta)$$

Denoting the initial value of ε by $[\varepsilon]$, to distinguish it from the variable quantity ε, then for $t=0$ we have $\varepsilon=[\varepsilon]$ and $\eta=0$, therefore for the same moment of time $p=0$, $q=[\varepsilon]$, whereby the equations (α) give $C=[\varepsilon]$, $D=0$; consequently by means of the equations (γ) and (δ)

$$\lambda = [\varepsilon]; \quad \frac{k}{2\,\omega}\lambda + \frac{\Omega}{\omega}\lambda' = 0$$

whence it follows that

$$\lambda' = -\frac{k}{2\Omega}[\varepsilon].$$

The values of C and D are therefore transformed into

$$C = [\varepsilon] \cos \Omega\,t; \quad D = -\frac{\omega}{\Omega}[\varepsilon] \sin \Omega\,t.$$

Now from the two equations (α) we find

$$p = -C\,e^{-\frac{1}{2}kt} \sin \omega t + D\,e^{-\frac{1}{2}kt} \cos \omega t$$
$$q = C\,e^{-\frac{1}{2}kt} \cos \omega t + D\,e^{-\frac{1}{2}kt} \sin \omega t$$

therefore

$$p = \varepsilon \sin \eta = -\tfrac{1}{2}[\varepsilon]\left(1 + \frac{\omega}{\Omega}\right) e^{-\frac{1}{2}kt} \sin (\omega + \Omega)\,t$$
$$-\tfrac{1}{2}[\varepsilon]\left(1 - \frac{\omega}{\Omega}\right) e^{-\frac{1}{2}kt} \sin (\omega - \Omega)\,t$$
$$q = \varepsilon \cos \eta = \tfrac{1}{2}[\varepsilon]\left(1 + \frac{\omega}{\Omega}\right) e^{-\frac{1}{2}kt} \cos (\omega + \Omega)\,t$$
$$+\tfrac{1}{2}[\varepsilon]\left(1 - \frac{\omega}{\Omega}\right) e^{-\frac{1}{2}kt} \cos (\omega - \Omega)\,t.$$

Since $\Omega^2 = \omega^2 - \frac{1}{4}k^2$, but k many times smaller than ω, we may in the integral assume $\omega = \Omega$, and therefore

$$\varepsilon \sin \eta = -[\varepsilon]\, e^{-\frac{1}{2}kt} \sin 2\,\omega t$$
$$\varepsilon \cos \eta = [\varepsilon]\, e^{-\frac{1}{2}kt} \cos 2\,\omega t$$

whence, by adding the squares and by division,

$$\varepsilon = [\varepsilon]\, e^{-\frac{1}{2}kt}; \quad \eta = 2\,\pi - 2\,\omega t \quad . \quad . \quad . \quad . \quad (\varepsilon)$$

Now, if $\varepsilon' = 0$,

$$\theta = \varepsilon \cos (\omega t + \eta)$$

therefore, by the substitution of the equations (ε),

$$\theta = [\varepsilon]\, e^{-\frac{1}{2}kt} \cos \omega t.$$

This expresses the following known theorem.

On the hypothesis that the resistance of the air is proportional to the velocity, the arcs of oscillation decrease in a geometrical proportion, while the time of oscillation remains the same.

Rigorously speaking, a very small decrease of the time of oscillation takes place, since $\Omega = \sqrt{\omega^2 - \frac{1}{4}k^2} < \omega$, but the difference is not perceptible.

The developements which hold good for the case $\varepsilon'=0$ may also be applied when ε' is very small, for in this case the variation of ε', arising from the resistance of the air, is a small quantity of the second order. The expression for $d\alpha$ in the equations (b) becomes on this supposition

$$\frac{d\alpha}{dt}=k\frac{\varepsilon'}{\varepsilon}\sin 2(\omega t+\eta)+\frac{k(\mu+\frac{1}{2}\mu')}{2\omega}\sin 2(\omega t+\eta)$$

and substituting in it the equations (ε),

$$\frac{d\alpha}{dt}=-\frac{k\varepsilon'}{[\varepsilon]}e^{\frac{1}{2}kt}\sin 2\omega t-\frac{k(\mu+\frac{1}{2}\mu')}{2\omega}\sin 2\omega t.$$

Since the second term of this equation gives after integrating much smaller terms than the first, we will omit it. Now

$$\int e^{\frac{1}{2}kt}\sin 2\omega t.dt=-\frac{8\omega e^{\frac{1}{2}kt}}{16\omega^2+k^2}\cos 2\omega t+\frac{2ke^{\frac{1}{2}kt}}{16\omega^2+k^2}\sin 2\omega t$$

of which again the second term is of a higher order than the first, and may therefore be neglected. Determining the arbitrary constant so that for $t=0$, also $\delta\alpha=0$, then

$$\delta\alpha=\frac{k\varepsilon'}{2\omega[\varepsilon]}(e^{\frac{1}{2}kt}\cos 2\omega t-1)$$

and hence at the moments of the greatest arcs of oscillation

$$\delta\alpha=\frac{k\varepsilon'}{2\omega[\varepsilon]}(e^{\frac{1}{2}kt}-1).$$

We have seen in the second approximation that the term which has a sensible effect on the motion of the plane of oscillation in the azimuth, is

$$\frac{d\alpha}{dt}=\tfrac{1}{24}\omega\varepsilon\varepsilon'$$

substituting herein for ε its value $[\varepsilon]\, e^{-\frac{1}{2}kt}$, and integrating, there results,

$$\delta\alpha = \frac{\omega\,[\varepsilon]\,\varepsilon'}{12\,k}\,(1 - e^{-\frac{3}{2}kt}).$$

This term is the greatest in the expression for the disturbance of the motion of the plane of oscillation in the azimuth.

We now proceed to investigate the effect of the resistance of the air on the motion of the pendulum if the latter moves on a conical surface of nearly circular base.

Substituting the additional values of D and E, viz.:

$$\begin{aligned} D &= k\,\omega\,(\varepsilon^2 - \varepsilon'^2)\sin(\omega t + \eta)\cos(\omega t + \eta) \\ E &= -k\,(\mu + \tfrac{1}{2}\mu')\,\theta^2 - k\,\omega\,\varepsilon\,\varepsilon' \end{aligned}$$

in the equations for the differentials of β, h, l, and α', or, what amounts to the same, substituting in the equations (b) for ε, ε', η, and α, and their differentials, the quantities β, h, l, and α', and their differentials, then we obtain

$$\begin{aligned} \frac{d\beta}{dt} &= -\tfrac{1}{2}\,k\,(\beta - l\cos 2\,\omega t + h\sin\omega t) \\ &\quad - k\,\frac{\mu + \frac{1}{2}\mu'}{2\,\omega}\,(\beta + l\cos 2\,\omega t - h\sin 2\,\omega t) \\ \frac{dh}{dt} &= -\frac{k}{2\,\beta}\,(\beta^2\sin 2\,\omega t + \beta h + l^2\sin 2\,\omega t + h\,l\cos 2\,\omega t) \\ &\quad - k\,\frac{\mu + \frac{1}{2}\mu'}{2\,\omega\,\beta}\,(\beta^2\sin 2\,\omega t - \beta h - l^2\sin 2\,\omega t - h\,l\cos 2\,\omega t) \\ \frac{dl}{dt} &= \frac{k}{2\,\beta}\,(\beta^2\cos 2\,\omega t - \beta l + h\,l\sin 2\,\omega t + h^2\cos 2\,\omega t) \\ &\quad + k\,\frac{\mu + \frac{1}{2}\mu'}{2\,\omega\,\beta}\,(\beta^2\cos 2\,\omega t + \beta l - h\,l\sin 2\,\omega t - h^2\cos 2\,\omega t) \\ \frac{d\alpha'}{dt} &= -\frac{k}{2\,\beta}\,(l\sin 2\,\omega t + h\cos 2\,\omega t) \\ &\quad + k\,\frac{\mu + \frac{1}{2}\mu'}{2\,\omega\,\beta}\,(l\sin 2\,\omega t + h\cos 2\,\omega t). \end{aligned}$$

In these equations the terms multiplied by $(\mu+\frac{1}{2}\mu')$ are much smaller than the others, they may therefore be neglected. Further, h and l are small quantities, on the supposition that the base of the conical surface is nearly a circle; therefore and because after integration the terms multiplied by h^2, $h\,l$, and l^2 have no small divisors, they will likewise be neglected. Consequently we have

$$\frac{d\beta}{dt}=-\tfrac{1}{2}k\beta+\tfrac{1}{2}k\,l\cos\omega\,t-\tfrac{1}{2}k\,h\sin 2\,\omega\,t$$

$$\frac{dh}{dt}=-\tfrac{1}{2}k\,\beta\sin 2\,\omega\,t-\tfrac{1}{2}k\,h$$

$$\frac{dl}{dt}=\tfrac{1}{2}k\,\beta\cos 2\,\omega\,t-\tfrac{1}{2}k\,l.$$

These equations are integrable, on the hypothesis that the resistance of the air is proportional to the velocity. Let

$$\beta=a\,e^{-\frac{1}{2}kt},\ h=a'\,e^{-\frac{1}{2}kt},\ l=a''\,e^{-\frac{1}{2}kt}$$

where e denotes the base of the hyperbolic logarithms, and a, a', a'' are functions of t. Substituting the differentials of these equations in the preceding, we have

$$\frac{da}{dt}=\tfrac{1}{2}k\,a''\cos 2\,\omega\,t-\tfrac{1}{2}k\,a'\sin 2\,\omega\,t$$

$$\frac{da'}{dt}=-\tfrac{1}{2}k\,a\sin 2\,\omega\,t$$

$$\frac{da''}{dt}=\tfrac{1}{2}k\,a\cos 2\,\omega\,t.$$

In order to obtain herefrom linear equations with constant coefficients, we assume

$$f=-a'\sin 2\,\omega\,t+a''\cos 2\,\omega\,t$$
$$g=\ \ a'\cos 2\,\omega\,t+a''\sin 2\,\omega\,t.$$

Therefore by differentiation, and with regard to the preceding equations,

$$\left.\begin{aligned} \frac{d\,a}{d\,t} &= \tfrac{1}{2}\,k\,f \\ \frac{d\,f}{d\,t} &= \tfrac{1}{2}\,k\,a - 2\,\omega g \\ \frac{d\,g}{d\,t} &= 2\,\omega\,f \end{aligned}\right\} \quad . \quad . \quad . \quad . \quad . \quad (\alpha\,\alpha)$$

The combination of the first and third of these equations gives by integration

$$a = \frac{k}{4\,\omega}\,g + \lambda \quad . \quad . \quad . \quad . \quad . \quad . \quad . \quad (\beta\,\beta)$$

where λ is the arbitrary constant, consequently the second equation is transformed into

$$\frac{d\,f}{d\,t} = \tfrac{1}{2}\,k\,\lambda - \frac{2}{\omega}\,(\omega^2 - \tfrac{1}{16}\,k^2)\,g.$$

The third equation, again differentiated, gives

$$\frac{d^2\,g}{d\,t^2} = 2\,\omega\,\frac{d\,f}{d\,t}$$

and this is transformed by the preceding into

$$\frac{d^2\,g}{d\,t^2} = -4\,(\omega^2 - \tfrac{1}{16}\,k^2)\,g + k\,\omega\,\lambda$$

the integral of which is

$$g = \lambda' \cos 2\,\Omega\,t + \lambda'' \sin 2\,\Omega\,t + \frac{k\,\omega\,\lambda}{4\,\Omega^2}$$

where λ' and λ'' are the arbitrary constants, and

$$\Omega = \sqrt{\omega - \tfrac{1}{16}\,k^2}.$$

From the third equation of $(\alpha\,\alpha)$ we derive

$$f = \frac{1}{2\,\omega}\,\frac{d\,g}{d\,t}$$

whence by differentiation and substitution

$$f = -\lambda' \frac{\Omega}{\omega} \sin 2\,\Omega\, t + \lambda'' \frac{\Omega}{\omega} \cos 2\,\Omega\, t \quad . \quad . \quad (\delta\,\delta)$$

In order to determine the arbitrary constants, we denote by $[\beta]$ the initial value of β, and by $[\beta']$ the initial value of β', that is, of $\frac{1}{2}\,(\varepsilon - \varepsilon')$. Since the initial value of η is zero, the initial value of h will likewise be zero, and $[\beta']$ will be the initial value of l. The same quantities are respectively the initial values of a, a', a'', whence it follows that $[\beta']$ is the initial value of f, and zero the initial value of g. Consequently the integrals give

$$\lambda = [\beta]\,;\ \lambda' = -\frac{k\,\omega\,[\beta]}{4\,\Omega^2}\,;\ \lambda'' = \frac{\omega\,[\beta']}{\Omega}.$$

Substituting these values, we obtain

$$a = \frac{k}{4\,\omega}\, g + [\beta]$$

$$g = -[\beta]\,\frac{k\,\omega}{4\,\Omega^2}\,\cos 2\,\Omega\, t + [\beta']\,\frac{\omega}{\Omega}\,\sin 2\,\Omega\, t + [\beta]\frac{k\,\omega}{4\,\Omega^2}$$

$$f = [\beta]\,\frac{k}{4\,\Omega}\,\sin 2\,\Omega\, t + [\beta']\cos 2\,\Omega\, t$$

and by means of our preceding assumptions

$$a' = -f \sin 2\,\omega\, t + g \cos 2\,\omega\, t$$

$$a'' = f \cos 2\,\omega\, t + g \sin 2\,\omega\, t$$

$$\beta = a\, e^{-\frac{1}{2}kt}\,;\ h = a'\, e^{-\frac{1}{2}kt}\,;\ l = a''\, e^{-\frac{1}{2}kt}\,;$$

therefore taking again $\Omega = \omega$, that is, neglecting k^2, we obtain

$$\beta = [\beta]\, e^{-\frac{1}{2}kt} + [\beta']\,\frac{k}{4\,\omega}\, e^{-\frac{1}{2}kt} \sin 2\,\omega\, t$$

$$h = -[\beta]\,\frac{k}{4\,\omega}\, e^{-\frac{1}{2}kt} + [\beta]\,\frac{k}{4\,\omega}\, e^{-\frac{1}{2}kt} \cos 2\,\omega\, t$$

$$l = [\beta']\, e^{-\frac{1}{2}kt} + [\beta]\,\frac{k}{4\,\omega}\, e^{-\frac{1}{2}kt} \sin 2\,\omega\, t.$$

Since $\beta' = \sqrt{h^2 + l^2}$, and $\tan 2\eta = \frac{h}{l}$, we have, neglecting k^2,

$$\beta' = [\beta'] e^{-\frac{1}{2}kt} + [\beta] \frac{k}{4\omega} e^{-\frac{1}{2}kt} \sin 2\omega t$$

$$\tan 2\eta = -\frac{[\beta] k}{4[\beta']\omega} + \frac{[\beta] k}{4[\beta']\omega} \cos 2\omega t.$$

At the moments of greatest amplitudes we have always $\cos 2\omega = 1$, hence the last formula shows that the resistance of the air does not affect the time of oscillation.

If the pendulum commences its motion on a conical surface with circular base, then $[\beta'] = 0$, and the preceding equations give

$$\beta = [\beta] e^{-\frac{1}{2}kt}$$

$$\beta' = [\beta] \frac{k}{4\omega} e^{-\frac{1}{2}kt} \sin 2\omega t.$$

The arcs of oscillation, therefore, decrease in a geometrical proportion, just as in the case when $\varepsilon' = 0$. Since $\beta' = 0$ at the points $\omega = 0$, $= 90°$, &c., we see that at these four points of a revolution the amplitude of the arcs of oscillation remains the same, so that the circular form of the base is preserved during the decrease of the arcs of oscillation, the deviation at the other points of revolution being imperceptible on account of the smallness of the factor k, which enters the expression for β'.

If the pendulum commences its motion on a conical surface with elliptical base, or $[\beta']$ is not equal to zero, then the equation for β' and β give by division

$$\frac{\beta'}{\beta} = \frac{[\beta'] + [\beta]\frac{k}{4\omega} \sin 2\omega t}{[\beta] + [\beta']\frac{k}{4\omega} \sin 2\omega t}$$

therefore at the four points above mentioned

$$\frac{\beta'}{\beta} = \frac{[\beta']}{[\beta]}$$

whence it follows that

$$\frac{\varepsilon'}{\varepsilon} = \frac{[\varepsilon']}{[\varepsilon]}$$

where $[\varepsilon]$ and $[\varepsilon']$ denote the initial semi-axes of the ellipse.

This equation proves the similarity of all the ellipses which the pendulum describes in consequence of the decrease of the arcs of oscillation caused by the resistance of the air.

Substituting the last values of β, h and l in the equation for $d\,\alpha'$, we obtain, neglecting again k^2,

$$\frac{d\,\alpha'}{d\,t} = -\frac{k\,[\beta']}{2\,[\beta]} \sin 2\,\omega\,t.$$

This equation shows that the time of rotation of the pendulum is not sensibly changed by the resistance of the air.

In order to justify our neglecting k^2, we will illustrate by a numerical example the smallness of the coefficient k in comparison with ω.

We have stated at the commencement of Chap. 7. Part I. the experiments made by General Dufour and Messrs. Wartmann and Marignac. The pendulum which they used was 20 metres in length, the amplitude of the first arc of oscillation amounted to $3^m, 25$, which was reduced, at the end of their experiments, to about $0^m, 70$. The duration of any of their experiments was upon the average $2^h, 243$. The least arcs of oscillation ε' were extremely small, and we have therefore here a case where $\varepsilon'=0$, whence the formula $\beta=[\beta]\,e^{-\frac{1}{2}kt}$ is transformed into

$$\varepsilon=[\varepsilon]\,e^{-\frac{1}{2}k\tau}$$

where $[\varepsilon]$ denotes the initial value of ε. Let $[\varepsilon_1]$ express the final value of ε, and τ the number of seconds elapsed between $[\varepsilon]$ and $[\varepsilon_1]$; then we have

$$[\varepsilon]=1^{\text{m}},625$$
$$[\varepsilon_1]=0^{\text{m}},350$$
$$\tau=2^{\text{h}},243\times 3600$$

and applying the above formula to this time and the two end-values of ε, we get

$$k=2\,\frac{\log[\varepsilon]-\log[\varepsilon_1]}{\text{M}\,\tau}$$

where M denotes the modulus of common logarithms.

The result of the numerical computation is

$$k=0{,}0003803.$$

But we have very nearly

$$\omega=\sqrt{\frac{g}{\lambda}}\,\cdot$$

and we may assume in this case the value of the accelerating force of gravity

$$g=9^{\text{m}},807$$

while the length of the pendulum is

$$\lambda=20^{\text{m}}$$

whence it follows that

$$\omega=0{,}7002145.$$

We see therefore that k, and the more so k^2, is an extremely small quantity in comparison to ω, and need only be retained where it enters an exponential function.

We will now examine the effect of the resisting medium on the conical motion of the free pendulum, on the other hypothesis that the resistance of the air is proportional to the square of the velocity.

The square of the angular velocity of the pendulum is, by equation (B_1)

$$\frac{d\theta^2}{dt^2}+\theta^2\frac{d\psi^2}{dt^2}$$

and we have therefore

$$k=\rho\sqrt{\frac{d\theta^2}{dt}+\theta^2\frac{d\psi^2}{dt^2}},$$

where ρ is a constant depending on the figure of the pendulum, and to be determined a posteriori, like k in our last example.

From the equations (p. 195)

$$\theta\frac{d\theta}{dt}=-\omega\,(\varepsilon^2-\varepsilon'^2)\sin(\omega t+\eta)\cos(\omega t+\eta)$$

$$\theta^2\frac{d\psi}{dt}=\omega\,\varepsilon\varepsilon'+(\mu+\tfrac{1}{2}\mu')\,\theta^2$$

we derive

$$\theta^2\left(\frac{d\theta^2}{dt^2}+\theta^2\frac{d\psi^2}{dt^2}\right)=\omega^2\,(\varepsilon^2-\varepsilon'^2)^2\sin^2(\omega t+\eta)\cos^2(\omega t+\eta)$$

$$+\omega^2\varepsilon^2\varepsilon'^2+2\,(\mu+\tfrac{1}{2}\mu')\,\omega\varepsilon\varepsilon'\theta^2+(\mu+\tfrac{1}{2}\mu')^2\,\theta^4$$

$$=\theta^3\,\omega^2\left[\varepsilon^2\sin^2(\omega t+\eta)+\varepsilon'^2\cos^2(\omega t+\eta)\right]$$

$$+2\,(\mu+\tfrac{1}{2}\mu')\,\omega\varepsilon\varepsilon'\,\theta^2+(\mu+\tfrac{1}{2}\mu')^2\,\theta^4.$$

Omitting the terms multiplied by $(\mu+\frac{1}{2}\mu')$, as they are much smaller than the others, we obtain

$$k=\rho\,\omega\sqrt{\quad\quad(\omega t+\eta)+\varepsilon'^2\cos^2(\omega t+\eta)}.$$

The square root in this expression can always be expanded

in a converging series, ascending in the cosines of multiples of $(\omega t+\eta)$, thus

$$k=\rho\omega\varepsilon\left[a^{(0)}+2a^{(1)}\cos 2(\omega t+\eta)+2a^{(2)}\cos 4(\omega t+\eta)+\&c.\right]$$

For the determination of the coefficients $a^{(0)}$, $a^{(1)}$, $a^{(2)}$, &c., we obtain, taking

$$c^2=\frac{\varepsilon^2-\varepsilon'^2}{\varepsilon^2}$$

the following expressions

$$a^{(0)}=\frac{1}{\pi}\int_0^\pi\sqrt{1-c^2\cos^2 x}\,.\,dx$$

$$a^{(1)}=\frac{1}{\pi}\int_0^\pi\sqrt{1-c^2\cos^2 x}\,.\,\cos 2x\,dx$$

&c.

which depend on the elliptic integrals the modulus of which is c, and they can therefore be calculated.

From the differential equations (b) we obtain by making herein $\mu+\frac{1}{2}\mu'=0$, and transforming the products and powers of the circular functions into linear dimensions

$$\frac{d\varepsilon}{dt}=-\tfrac{1}{2}k\varepsilon+\tfrac{1}{2}k\varepsilon\cos 2(\omega t+\eta)$$

$$\frac{d\varepsilon'}{dt}=-\tfrac{1}{2}k\varepsilon'-\tfrac{1}{2}k\varepsilon'\cos 2(\omega t+\eta)$$

$$\frac{d\eta}{dt}=-\tfrac{1}{2}k\frac{\varepsilon^2+\varepsilon'^2}{\varepsilon^2-\varepsilon'^2}\sin 2(\omega t+\eta)$$

$$\frac{d\alpha}{dt}=k\frac{\varepsilon\varepsilon'}{\varepsilon^2-\varepsilon'^2}\sin 2(\omega t+\eta).$$

Substituting the series for k in these equations, they become

$$\frac{d\epsilon}{dt}=-\tfrac{1}{2}\rho\omega\epsilon^2(a^{(0)}-a^{(1)})-\tfrac{1}{2}\rho\omega\epsilon^2(2a^{(1)}-a^{(0)}-a^{(2)})\cos 2(\omega t+\eta)$$
$$-\tfrac{1}{2}\rho\omega\epsilon^2(2a^{(2)}-a^{(1)}-a^{(3)})\cos 4(\omega t+\eta)-\&c.$$

$$\frac{\delta \epsilon'}{d t} = -\tfrac{1}{2}\rho\omega\epsilon\epsilon'(a^{(0)} + a^{(1)}) - \tfrac{1}{2}\rho\omega\epsilon\epsilon'(2a^{(1)} + a^{(0)} + a^{(3)})\cos 2(\omega t + \eta)$$
$$-\tfrac{1}{2}\rho\omega\epsilon\epsilon'(2a^{(2)} + a^{(1)} + a^{(3)})\cos 4(\omega t + \eta) - \&c.$$
$$\frac{\eta}{d t} = -\tfrac{1}{2}\rho\omega\epsilon\frac{\epsilon^2 + \epsilon'^2}{\epsilon^2 - \epsilon'^2}(a^{(0)} - a^{(2)})\sin 2(\omega t + \eta)$$
$$-\tfrac{1}{2}\rho\omega\epsilon\frac{\epsilon^2 + \epsilon'^2}{\epsilon^2 - \epsilon'^2}(a^{(1)} - a^{(3)})\sin 4(\omega t + \eta) - \&c.$$
$$\frac{\delta \alpha}{d t} = \rho\omega\frac{\epsilon^2\epsilon'}{\epsilon^2 - \epsilon'^2}(a^{(0)} - a^{(2)})\sin 2(\omega t + \eta)$$
$$+\rho\omega\frac{\epsilon^2\epsilon}{\epsilon^2 - \epsilon'^2}(a^{(1)} - a^{(3)})\sin 4(\omega t + \eta) + \&c.$$

Taking on the right-hand side of these equations $\eta = 0$, their integration gives the following results:

$$\delta\varepsilon = -\tfrac{1}{2}\rho\varepsilon^2(a^{(0)} - a^{(1)})\omega t - \tfrac{1}{4}\rho\varepsilon^2(2a^{(1)} - a^{(0)} - a^{(2)})\sin 2\omega t$$
$$-\tfrac{1}{8}\rho\varepsilon^2(2a^{(2)} - a^{(1)} - a^{(3)})\sin 4\omega t - \&c.$$
$$\delta\varepsilon' = -\tfrac{1}{2}\rho\varepsilon\varepsilon'(a^{(0)} + a^{(1)})\omega t - \tfrac{1}{4}\rho\varepsilon\varepsilon'(2a^{(1)} + a^{(0)} + a^{(2)})\sin 2\omega t$$
$$-\tfrac{1}{8}\rho\varepsilon\varepsilon'(2a^{(2)} + a^{(1)} + a^{(3)})\sin 4\omega t - \&c.$$
$$\delta\eta = \tfrac{1}{4}\rho\varepsilon\frac{\varepsilon^2 + \varepsilon'^2}{\varepsilon^2 - \varepsilon'^2}(a^{(0)} - a^{(2)})\cos 2\omega t$$
$$+\tfrac{1}{8}\rho\varepsilon\frac{\varepsilon^2 + \varepsilon'^2}{\varepsilon^2 - \varepsilon'^2}(a^{(1)} - a^{(3)})\cos 4\omega t + \&c.$$
$$\delta\alpha = -\tfrac{1}{2}\rho\frac{\varepsilon^2\varepsilon'}{\varepsilon^2 - \varepsilon'^2}(a^{(0)} - a^{(2)})\cos 2\omega t$$
$$-\tfrac{1}{4}\rho\frac{\varepsilon^2\varepsilon'}{\varepsilon^2 - \varepsilon'^2}(a^{(1)} - a^{(3)})\cos 4\omega t - \&c.$$

Taking these integrals between the limits $\omega t = 0$ and $\omega t = \pi$, we obtain firstly

$$\delta\eta = 0;\ \delta\alpha = 0$$

which shows that the resistance of the air has neither an influence on the time of oscillation nor on the azimuth of the plane of oscillation.

We obtain secondly

$$\delta\varepsilon = -\tfrac{1}{2}\rho\,\varepsilon^2\,(a^{(0)} - a^{(1)})\,\pi\,;\quad \delta\varepsilon' = -\tfrac{1}{2}\rho\,\varepsilon\,\varepsilon'\,(a^{(0)} + a^{(1)})\,\pi$$

which equations express the decrease of the greatest and smallest amplitude during one oscillation.

Since this decrease is only very small, it may be extended over any number of oscillations in the following way.

Let us denote the initial greatest and least amplitude by ε and ε'; the same after the lapse of one oscillation by ε_1 and ε'_1; the same after the lapse of two oscillations by ε_2 and ε'_2 and so on; further, let the respective coefficients of ε_1 and ε'_1, of ε_2 and ε'_2, &c., be denoted by $a_1^{(0)}$ and $a_1^{(1)}$ by $a_2^{(0)}$ and $a_2^{(1)}$, &c., then we have with sufficient exactitude.

$$\left.\begin{aligned}
\varepsilon_1 &= \varepsilon - \tfrac{1}{2}\rho\,\varepsilon^2\,(a^{(0)} - a^{(1)})\,\pi \\
\varepsilon'_1 &= \varepsilon' - \tfrac{1}{2}\rho\,\varepsilon\,\varepsilon'\,(a^{(0)} + a^{(1)})\,\pi \\
\varepsilon_2 &= \varepsilon_1 - \tfrac{1}{2}\rho\,\varepsilon_1^2\,(a_1^{(0)} - a_1^{(1)})\,\pi \\
\varepsilon'_2 &= \varepsilon'_1 - \tfrac{1}{2}\rho\,\varepsilon_1\varepsilon'_1\,(a_1^{(0)} + a_1^{(1)})\,\pi \\
\varepsilon_3 &= \varepsilon_2 - \tfrac{1}{2}\rho\,\varepsilon_2^2\,(a_2^{(0)} - a_2^{(1)})\,\pi \\
\varepsilon'_3 &= \varepsilon'_2 - \tfrac{1}{2}\rho\,\varepsilon_2\varepsilon'_2\,(a_2^{(0)} + a_2^{(1)})\,\pi
\end{aligned}\right\}\ (Z)$$

&c.

These equations may be continued to the last observed oscillation, and by means of them the value of ρ can be calculated.

Considering the case, when $\varepsilon' = 0$, we have

$$a^{(0)} = \frac{1}{\pi}\int_0^\pi \sin x\,.\,dx = \frac{2}{\pi}$$

$$a^{(1)} = \frac{1}{\pi}\int_0^\pi \sin x\cos 2x\,dx = -\frac{2}{1\,.\,3\,.\,\pi}$$

$$a^{(2)} = \frac{1}{\pi}\int_0^\pi \sin x\cos 4x\,dx = -\frac{2}{3\,.\,5\,.\,\pi}$$

$$a^{(3)} = \frac{1}{\pi}\int_0^\pi \sin x\cos 6x\,dx = -\frac{2}{5\,.\,7\,.\,\pi}$$

&c.

whence we derive the series

$$\sin x = \frac{2}{\pi} - 2\frac{2}{1.3.\pi}\cos 2x - 2\frac{2}{3.5.\pi}\cos 4x - 2\frac{2}{5.7.\pi}\cos 6x - \&c.$$

The convergency of this series will depend on the convergency of the sum of the coefficients, but

$$\frac{1}{1.3} + \frac{1}{3.5} + \frac{1}{5.7} + \&c. = \tfrac{1}{2}\left[\begin{array}{l} 1 + \tfrac{1}{3} + \tfrac{1}{5} + \tfrac{1}{7} + \&c. \\ -(\tfrac{1}{3} + \tfrac{1}{5} + \tfrac{1}{7} + \&c.) \end{array}\right]$$
$$= \tfrac{1}{2}$$

hence our series for $\sin x$ is converging at least for the interval from $x = 0$ to $x = \pi$; and it is between these limits we must take the variable quantity, since we apply our formulæ to each oscillation separately. Now the condition that $\varepsilon' = 0$ gives $\delta\varepsilon' = 0$, and if we substitute the values

$$a^{(0)} = \frac{2}{\pi};\ a^{(1)} = -\frac{2}{3\pi}$$

in the equations (Z), we obtain

$$\begin{aligned} \varepsilon_1 &= \varepsilon - \tfrac{4}{3}\rho\varepsilon^2 \\ \varepsilon_2 &= \varepsilon_1 - \tfrac{4}{3}\rho\varepsilon_1^2 \\ \varepsilon_3 &= \varepsilon_2 - \tfrac{4}{3}\rho\varepsilon_2^2 \\ &\&c. \end{aligned}$$

a result agreeable to the well known theorem on the course of the pendulum in resisting medium.

We may proceed a step farther. If we suppose ε' very small, the corresponding values of $a^{(0)}$ and $a^{(1)}$ differ from those just derived only by a quantity of the order $\left(\frac{\varepsilon'}{\varepsilon}\right)^2$, and we may therefore assume that the preceding values hold good when ε' is small. Substituting these values in the equations (Z), we have

$$\varepsilon'_1 = \varepsilon' - \tfrac{2}{3}\rho\,\varepsilon\,\varepsilon'$$
$$\varepsilon'_2 = \varepsilon'_1 - \tfrac{2}{3}\rho\,\varepsilon_1\,\varepsilon'_1$$
$$\varepsilon'_3 = \varepsilon'_2 - \tfrac{2}{3}\rho\,\varepsilon_2\,\varepsilon'_2$$
&c.

We derive therefrom

$$\frac{\varepsilon_\mu}{\varepsilon_{\mu-1}} = 1 - \tfrac{4}{3}\rho\,\varepsilon_{\mu-1}; \quad \frac{\varepsilon'_\mu}{\varepsilon'_{\mu-1}} = 1 - \tfrac{2}{3}\rho\,\varepsilon_{\mu-1}$$

therefore the ratio of any two consecutive greatest amplitudes of oscillation differs twice as much from unity as the two corresponding smallest amplitudes, a result which is essentially at variance with that found on the hypothesis, that the resistance of the air is proportional to the velocity itself.

Let us finally consider the case when the base of the conical surface, on which the pendulum moves, is nearly circular.

Introducing into the expression for k the quantities

$$\beta = \tfrac{1}{2}(\varepsilon + \varepsilon'); \quad \beta' = \tfrac{1}{2}(\varepsilon - \varepsilon'); \quad h = \beta' \sin 2\eta;$$
$$l = \beta' \cos 2\eta; \quad \alpha' = \alpha + \eta,$$

we have

$$k = \rho\,\omega\,\sqrt{\beta^2 + \beta'^2 - 2\beta(l\cos 2\omega t - h \sin 2\omega t)}$$

or, only considering the first power of β',

$$k = \rho\,\omega\,\beta - \rho\,\omega\,(l\cos 2\omega t - h\sin 2\omega t).$$

From the differential equations for β, h, l, α' we get, by making $\mu + \tfrac{1}{2}\mu' = 0$, and neglecting the powers and products of h and l,

$$\frac{d\beta}{dt} = -\tfrac{1}{2}k(\beta - l\cos 2\omega t + h\sin 2\omega t)$$

$$\frac{dh}{dt} = -\tfrac{1}{2}k\,\beta\sin 2\omega t - \tfrac{1}{2}k\,h$$

$$\frac{dl}{dt} = \tfrac{1}{2} k \beta \cos 2 \omega t - \tfrac{1}{2} k l$$
$$\frac{d\alpha'}{dt} = -\tfrac{1}{2} \frac{k}{\beta}(l \sin 2 \omega t + h \cos 2 \omega t)$$

whence, by means of the preceding value of k, we obtain the following differential equations —

$$\frac{d\beta}{dt} = -\tfrac{1}{2} \rho \omega \beta^2 + \rho \omega \beta (l \cos 2 \omega t - h \sin \omega t)$$
$$\frac{dh}{dt} = -\tfrac{1}{2} \rho \omega \beta^2 \sin 2 \omega t - \tfrac{3}{4} \rho \omega \beta h$$
$$+ \tfrac{1}{4} \rho \omega \beta (l \sin 4 \omega t + h \cos 4 \omega t)$$
$$\frac{dl}{dt} = \tfrac{1}{2} \rho \omega \beta^2 \cos 2 \omega t - \tfrac{3}{4} \rho \omega \beta l$$
$$- \tfrac{1}{4} \rho \omega \beta (l \cos 4 \omega t - h \sin 4 \omega t)$$
$$\frac{d\alpha'}{dt} = -\tfrac{1}{2} \rho \omega (l \sin 2 \omega t + h \cos 2 \omega t).$$

Integrating these equations from $\omega t = 0$ to $\omega t = \pi$, that is, for the time of one oscillation, there results

$$\delta \beta = -\tfrac{1}{2} \pi \rho \beta^2$$
$$\delta h = -\tfrac{3}{4} \pi \rho \beta h$$
$$\delta l = -\tfrac{3}{4} \pi \rho \beta l$$
$$\delta \alpha' = 0.$$

But the initial value of η is $\eta = 0$, and therefore

$$h = 0, \quad l = \beta'$$

whence the second integral becomes

$$\delta h = 0$$

and, because $\frac{h}{l} = \tan 2 \eta$, it follows that

$$\delta \eta = 0.$$

This integral, in conjunction with $\delta\alpha'=0$, shows that at the moments of greatest amplitudes the resistance of the air has neither an influence on the time of oscillation, nor on the azimuth of the plane of oscillation.

Since for the initial value $\eta=0$, we have $l=\beta'$, therefore the integral

$$\delta l=-\tfrac{3}{4}\pi\rho\beta l$$

becomes

$$\delta\beta'=-\tfrac{3}{4}\pi\rho\beta\beta'$$

and making use of a similar notation as before, we obtain

$$\begin{aligned}
\beta_1&=\beta-\tfrac{1}{2}\pi\rho\beta^2; & \beta_1'&=\beta'-\tfrac{3}{4}\pi\rho\beta\beta'\\
\beta_2&=\beta_1-\tfrac{1}{2}\pi\rho\beta_1^2; & \beta_2'&=\beta_1'-\tfrac{3}{4}\pi\rho\beta_1\beta_1'\\
\beta_3&=\beta_2-\tfrac{1}{2}\pi\rho\beta_2^2; & \beta_3'&=\beta_2-\tfrac{3}{4}\pi\rho\beta_2\beta_2'.
\end{aligned}$$

These expressions show that $(\epsilon-\epsilon')$ being small, the ratio of ϵ to ϵ' undergoes changes during the motion of the pendulum, a result which differs again from that found on the hypothesis that the resistance varies as the velocity itself.

The torsion of the thread.—The force of torsion is proportional to the angle of torsion, which in our case is equal to the angle through which the pendulum's bob turns in the rotation about its axis, and is therefore by our preceding notation $\int r\,dt$. Let k denote the momentum of torsion for the angular unit in circular measure, then, because the rotation of the pendulum is opposed by the force of torsion, the momentum of the latter will be in general expressed by

$$-k\int r\,dt.$$

This is the term to be added to the right-hand side of the first equation of (L), so that, taking again $A=B$, we have, with regard to the correction on account of the resistance to twisting,

$$C\frac{dr}{dt}=-k\int r\,dt$$

taking for shortness

$$R=\int r\,dt$$

we have

$$\frac{d^2 R}{dt^2}=-\frac{kR}{C}$$

the integral of which is

$$R=c\cos\omega' t+c'\omega' t$$

where

$$\omega'=\sqrt{\frac{k}{C}}$$

and where c and c' denote the two arbitrary constants which are to be determined by the initial values of R and $\frac{dR}{dt}$.

Let the initial angular velocity of the pendulum's rotation on its axis be denoted as in (A_1), viz., by n', and suppose that for the time $t=0$, the torsion is $-\tau$; then

$$c=-\tau;\ \omega' c'=n'$$

and we obtain

$$R=-\tau\cos\omega' t+\frac{n'}{\omega'}\sin\omega' t.$$

This equation shows that the rotation of the pendulum on its axis is changed by the force of torsion into an oscillating motion, the greatest amplitude of which is

$$\pm\sqrt{\tau^2+\frac{n'^2}{\omega'^2}}.$$

The same expression gives by differentiation, since $r = \frac{dR}{dt}$, the following:

$$r = \omega' \tau \sin \omega' t + n' \cos \omega' t.$$

This is the quantity which must be substituted for the value of n' in the equation (F_1), where $\mu' = n' \frac{C}{A}$, so that now the value of μ' becomes

$$\mu' = \frac{C \omega' \tau}{A} \sin \omega' t + \frac{C n'}{A} \cos \omega' t.$$

We will now proceed to investigate the variation of the arbitrary constants arising from this value of μ'.

From the equation (C_2) we have, neglecting μ,

$$v = \omega \frac{\varepsilon'}{\varepsilon} + \tfrac{1}{2} \mu'$$

whence by squaring and omitting μ'^2,

$$v^2 \varepsilon^2 = \omega^2 \varepsilon'^2 + \mu' \omega \varepsilon \varepsilon'.$$

Taking as before $\sin \varepsilon = \varepsilon$, the first equation (G_1) gives, with omission of the constants and those terms multiplied by $\mu \mu'$,

$$\kappa = \mu' \omega \varepsilon \varepsilon'$$

and the second equation gives on the same conditions $\kappa' = 0$, so that only that term of the second equation (F_1) which is multiplied by μ' enters the value of κ', viz.,

$$\kappa' = \tfrac{1}{2} \mu' \theta^2.$$

Substituting these values of κ and κ' in the first system of differential equations for ε, ε', η, and α, then their integrals will express the additional effect of torsion on the motion of the pendulum. But before making this substitution, we must derive the following expressions.

By differentiating the equations for κ and κ' we have

$$\frac{d\kappa}{dt} = \frac{d\mu'}{dt} - \omega\,\varepsilon\,\varepsilon'$$

$$\frac{d\kappa'}{dt} = \tfrac{1}{2}\,\frac{d\mu'}{dt}\,\theta^2 + \mu'\,\theta\,\frac{d\theta}{dt}.$$

Since we must here take

$$\frac{d\psi}{dt} = \frac{\omega\,\varepsilon\,\varepsilon'}{\theta^2}$$

we get

$$\tfrac{1}{2}\,\theta\,\frac{d\kappa}{dt} = \tfrac{1}{2}\,\frac{d\mu'}{dt}\,\theta\,\omega\,\varepsilon\,\varepsilon'$$

$$\theta\,\frac{d\psi}{dt}\,\frac{d\kappa'}{dt} = \tfrac{1}{2}\,\frac{d\mu'}{dt}\,\theta\,\omega\,\varepsilon\,\varepsilon' + \mu'\,\frac{d\theta}{dt}\,\omega\,\varepsilon\,\varepsilon'.$$

By means of these values the expressions for the auxiliary quantities D and E give

$$D = -\mu'\,\omega\,\varepsilon\,\varepsilon'$$

$$E = \mu'\,\theta\,\frac{d\theta}{dt} + \tfrac{1}{2}\,\frac{d\mu'}{dt}\,\theta^2$$

whence the differential equations are transformed into

$$\frac{d\epsilon}{dt} = \tfrac{1}{2}\,\mu'\,\epsilon'\sin 2\,(\omega t + \eta) + \frac{\epsilon'}{4\,\omega}\,\frac{d\mu'}{dt}\,[1 - \cos 2\,(\omega t + \eta)]$$

$$\frac{d\epsilon'}{dt} = -\tfrac{1}{2}\,\mu'\,\epsilon\sin 2\,(\omega t + \eta) + \frac{\epsilon}{4\,\omega}\,\frac{d\mu'}{dt}\,[1 + \cos 2\,(\omega t + \eta)]$$

$$\frac{d\eta}{dt} = \mu'\,\frac{\epsilon\,\epsilon'}{\epsilon^2 - \epsilon'^2}\cos 2\,(\omega t + \eta) + \frac{\epsilon\,\epsilon'}{2\,\omega\,(\epsilon^2 - \epsilon'^2)}\,\frac{d\mu'}{dt}\sin 2\,(\omega t + \eta)$$

$$\frac{d\alpha}{dt} = \tfrac{1}{2}\,\mu' - \tfrac{1}{2}\,\mu'\,\frac{\epsilon^2 + \epsilon'^2}{\epsilon^2 - \epsilon'^2}\cos 2\,(\omega t + \eta) - \frac{\epsilon^2 + \epsilon'^2}{4\,\omega\,(\epsilon^2 - \epsilon'^2)}\,\frac{d\mu'}{dt}\sin 2\,(\omega t + \eta$$

where, according to a preceding result,

$$\mu' = \frac{C\,\omega'\,\tau}{A}\sin\omega' t + \frac{C\,n'}{A}\cos\omega' t.$$

The value of $\omega' = \sqrt{\frac{k}{C}}$ depends chiefly on the intensity k of the force of torsion, and the latter again depends on the cross-section of the thread. But we are always able to choose the thickness of the thread so that ω' may be many times smaller than ω; and on this supposition the preceding differential equations show that by the integration no terms can arise which are multiplied into the time, and more, that such terms can only be then generated when $\omega' = 2\omega$ contrary to our supposition. It is further evident that of all the terms arising from the integration those must be the greatest which have the small divisor ω'. But these latter terms can only be generated from the first term of the equation for $\delta\alpha$.

Therefore we take

$$\frac{d\alpha}{dt} = \tfrac{1}{2}\mu' = \tfrac{1}{2}\frac{C\omega'\tau}{A}\sin\omega' t + \tfrac{1}{2}\frac{Cn'}{A}\cos\omega' t.$$

The integration of this equation gives by determining the constant, so that for $t=0$, also $\delta\alpha=0$, the following result,

$$\delta\alpha = \frac{C\tau}{A}\sin^2\tfrac{1}{2}\omega' t + \frac{Cn'}{2\omega' A}\sin\omega' t.$$

Of all the expressions arising from the torsion of the thread this is the only one which may become perceptible. Its maximum effect produced on the change of the azimuth of the plane of oscillation is

$$\frac{C\tau}{2A} + \frac{C}{2A}\sqrt{\tau^2 + \frac{n'^2}{\omega'^2}}$$

and differs therefore essentially from that produced by the earth's rotation, which has no maximum.

If the pendulum is at perfect rest before its judicious liberation, the initial torsion will be very small and may be even zero. In this case the maximum of the change amounts to

$$\frac{C\,n'}{2\,\omega'\,A}.$$

In order that this quantity may not become great, we must take care that the velocity n' be as small as possible.

For illustration we take the numerical example given in our first approximation (p. 188), where we found

$$\tfrac{1}{2}\,\mu'\,T = 162''\,u.$$

But in this expression we have

$$\mu' = \frac{C\,n'}{A}; \quad T = \frac{\pi}{\omega}$$

therefore by substitution

$$\frac{C\,n'}{2\,A} = \frac{162''\,u\,\omega}{\pi}.$$

Let us take the case where $2\,u = 1$, and $100\,\omega' = \omega$, then

$$\frac{C\,n'}{2\,\omega'\,A} = \frac{8100''}{\pi} = 43 \text{ minutes}$$

which is a quantity disturbing the motion of the pendulum very considerably.

It is therefore advisable to relinquish Mr. Faucault's mode of suspension, and to adopt one or the other described in Chap. 7. Part I., where the thickness of the suspending wire excludes all possibility of its torsion.

CHAPTER II.

THEORY OF THE GYROSCOPE.

THE theory of the gyroscope may be founded upon the theorems concerning the apparent forces in relative motion; but these theorems not being generally known the author * has preferred to base his researches upon the consideration of absolute motion.

The principal parts of the gyroscope (*fig.* 29) are a ring

FIG. 29.

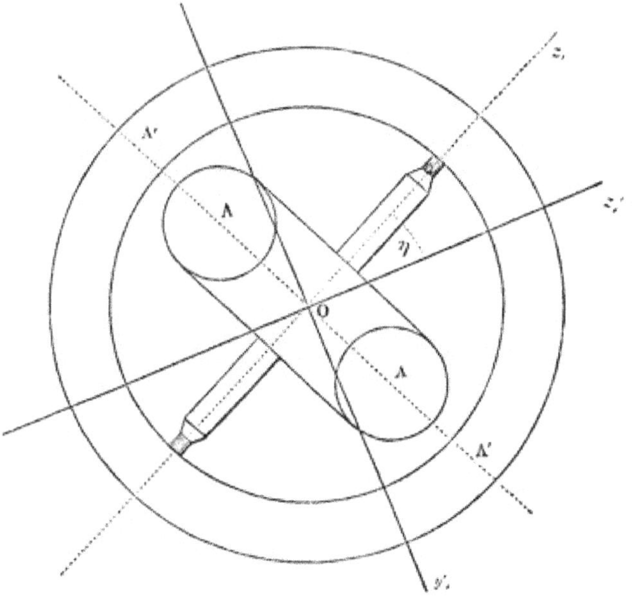

* M. Yvon Villarceau.

A′ movable about a fixed axis $Oz_{,}'$ within which a solid of revolution A turns about its axis $Oz_{,}$ which is constrained in the motion of the ring. Let the axes of the ring and disc, intersecting in the common centre of gravity O, have any inclination to each other denoted by η, but let us admit that they are principal axes respectively. One of the three rectangular axes x, y, z, must always be made to coincide with an axis of rotation by which means the positive direction of rotation and of moments is easily determined.

Describe with centre O a sphere, and through the same point a fixed plane, which cuts the sphere along the great circle NN′xy (*fig.* 30). Draw in this plane two rect-

Fig. 30.

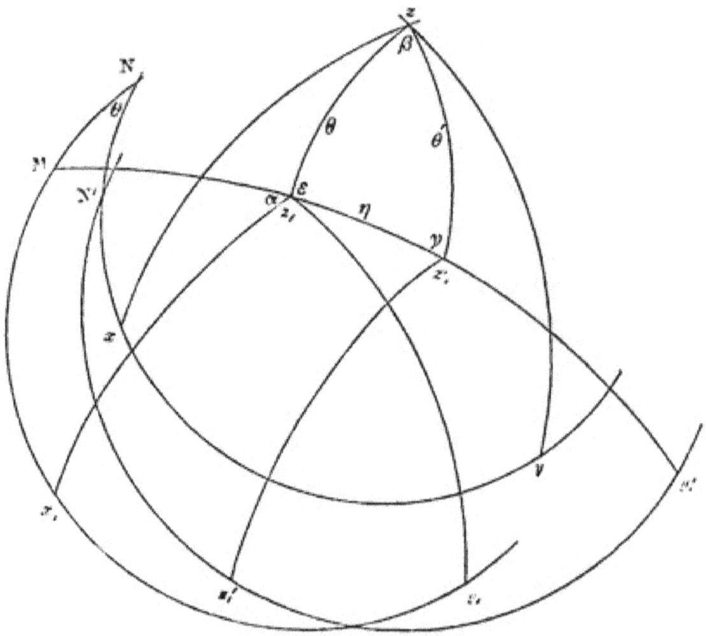

angular axes x and y, and perpendicular to it an axis z. Let us assume the fixed plane parallel to the terrestrial equator which we suppose to be immovable during the duration of the experiment. The points x, y, z, where these axes intersect the sphere, are therefore: the first, the vernal equinox; the other, 90° farther in the direction of the earth's rotation; the third, the north pole. In each of the equatorial planes of the two movable axes $z_{,}$ and $z_{,}'$ draw two other movable axes perpendicular to each other and denoted by $x_{,}, y_{,}$ and $x_{,}', y_{,}'$. The equatorial plane of the disc cuts the terrestrial equator in a point N distant from the axis of x by an angle $\mathrm{N}x = \psi$, and makes with the plane of $x\,y$ an angle $=\theta$ on the negative side of the axis of z. Denoting by ϕ the distance of the axis of $x_{,}$ from the point N, we have $\mathrm{N}\,y_{,} = 90 + \phi$. The pole of the equator of the disc is in $z_{,}$, and its distance from the axis of z is measured by the angle θ which always lies between zero and 180°, while the angles ψ and ϕ have no limits.

The same letters with accents express the same quantities in relation to the ring. The polar distance $z\,z_{,}' = \theta'$ is constant; the plane joining the poles z and $z_{,}'$ is constrained in the diurnal motion, and the spherical angle $x\,z\,z_{,}'$ is the right ascension R of the axis $z_{,}'$ of the ring. Denoting by ω the angular velocity of the earth's motion, and by $d\,t$ the element of time we have

$$\frac{d\mathrm{R}}{dt} = \omega \quad . \quad . \quad . \quad . \quad . \quad . \quad . \quad . \quad . \quad (1)$$

For brevity's sake we term the plane passing through the axes of the ring and disc, the plane of the ring, and that passing through the axes of the ring and earth, the horary plane, denoting by γ the angle between these two planes.

In order to determine the position of the disc with regard to the ring, we produce the plane of the latter to meet the equator of the former, in M suppose, denoting by α the angle $\mathrm{M}z_{,}x_{,}$. It is now our object to express the circular quantities α and γ, which determine the positions of the disc and ring in functions of the time.

Let us first find expressions for the position of the movable axes with regard to each other:

$x_{,}'$ being the pole of $\mathrm{M}\,z_{,}z_{,}'$, the arc joining $z_{,}$ and $x_{,}'$ will be perpendicular to it, therefore $x_{,}z_{,}x_{,}'=90^{\circ}-\alpha$, whence

$$\cos(x_{,}x_{,}')=\sin\alpha \quad . \quad . \quad . \quad . \quad . \quad . \quad . \quad . \quad . \quad (2)$$

Joining $x_{,}$ and $y_{,}'$ the triangle $x_{,}z_{,}y_{,}'$ where the angle at $z'=180^{\circ}-\alpha$, will give, because the side $z_{,}y'_{,}=90^{\circ}+\eta$,

$$\cos(x_{,}y_{,}')=-\cos\eta\cos\alpha \quad . \quad . \quad . \quad . \quad . \quad . \quad (3)$$

In the same manner we obtain from the triangle $x_{,}z_{,}z_{,}'$

$$\cos(x_{,}z_{,}')=-\sin\eta\cos\alpha \quad . \quad . \quad . \quad . \quad . \quad . \quad (4)$$

Completing the triangle $x_{,}'z_{,}y_{,}$ and remembering that the angle at $z_{,}=\alpha$, we shall have,

$$\cos(y_{,}x_{,}')=\cos\alpha \quad . \quad . \quad . \quad . \quad . \quad . \quad . \quad . \quad . \quad (5)$$

and from the triangle $y_{,}z_{,}y_{,}'$ where angle at $z_{,}=90^{\circ}-\alpha$.

$$\cos(y_{,}y_{,}')=\cos\eta\sin\alpha \quad . \quad . \quad . \quad . \quad . \quad . \quad . \quad (6)$$

Lastly, from the triangle $y_{,}z_{,}z_{,}'$ where likewise angle at $z_{,}=90^{\circ}-\alpha$.

$$\cos(y_{,}z_{,}')=\sin\eta\sin\alpha \quad . \quad . \quad . \quad . \quad . \quad . \quad . \quad (7)$$

Besides, we have directly,

$$\left.\begin{array}{l}\cos(z, x_{\prime}')=0\\ \cos(z, y_{\prime}')=-\sin\eta\\ \cos(z, z_{\prime}')=\cos\eta.\end{array}\right\} \quad . \quad . \quad . \quad . \quad (8)$$

The arcs N z and N $z_{\prime}$ being quadrants, the angles at z and $z_{\prime}$ in triangle N $z z_{\prime}$ are right angles, and therefore the sum N $z x + x z z_{\prime} = 90°$, or denoting by β the angle $z_{\prime} z z'$,

$$\psi + \mathrm{R} - \beta = 90° \quad . \quad . \quad . \quad . \quad . \quad . \quad . \quad . \quad (8 \text{ bis})$$

and substituting ε for angle $z\, z_{\prime} z_{\prime}'$ we have $\varepsilon + \mathrm{N} z_{\prime} \mathrm{M} = 90°$, or

$$\phi - \alpha = 90° - \varepsilon \quad . \quad . \quad . \quad . \quad . \quad . \quad . \quad . \quad . \quad . \quad . \quad (9)$$

The triangle N$'z\, z_{\prime}'$, having likewise two right angles, gives firstly

$$\psi' = 90° - \mathrm{R} \quad . \quad . \quad . \quad . \quad . \quad . \quad . \quad . \quad . \quad . \quad . \quad (10)$$

then $\gamma + \mathrm{M}\, z_{\prime}' \mathrm{N}' = 90°$; but because $\mathrm{M}\, z_{\prime}' \mathrm{N}' = 90° - \phi'$, it follows

$$\gamma = \phi' \quad . \quad . \quad . \quad . \quad . \quad . \quad . \quad . \quad . \quad . \quad . \quad . \quad . \quad (11)$$

The value of ψ from equation (8 bis)

$$\psi = 90° - \mathrm{R} + \beta \quad . \quad . \quad . \quad . \quad . \quad . \quad . \quad . \quad . \quad (12)$$

compared with that of ψ' in (10) gives

$$\psi - \psi' = \beta \quad . \quad . \quad . \quad . \quad . \quad . \quad . \quad . \quad . \quad . \quad (13)$$

The following systems of equations are derived from the properties of the spherical triangle $z\, z_{\prime} z_{\prime}'$:

$$\left.\begin{array}{l}\sin\beta\sin\theta = \sin\eta\sin\gamma\\ \sin\theta'\sin\gamma = \sin\theta\sin\varepsilon\end{array}\right\} \quad . \quad . \quad . \quad (14)$$

$$\left.\begin{array}{l}\cos\eta\cos\theta' + \sin\eta\sin\theta'\cos\gamma = \cos\theta\\ \cos\eta\cos\theta + \sin\eta\sin\theta\cos\varepsilon = \cos\theta'\\ \cos\theta'\cos\theta + \sin\theta'\sin\theta\cos\beta = \cos\eta\end{array}\right\} \quad . \quad . \quad . \quad (15)$$

$$\left.\begin{aligned} -\cos\gamma\cos\varepsilon+\sin\gamma\sin\varepsilon\cos\eta&=\cos\beta \\ -\cos\beta\cos\gamma+\sin\beta\sin\gamma\cos\theta'&=\cos\varepsilon \end{aligned}\right\} \quad . \quad . \quad . \quad (16)$$

$$\left.\begin{aligned} \sin\theta'\cos\beta+\sin\eta\cos\theta\cos\varepsilon&=\cos\eta\sin\theta \\ \sin\theta'\cos\gamma+\sin\theta\cos\eta\cos\varepsilon&=\cos\theta\sin\eta \\ \sin\theta\cos\varepsilon+\sin\theta'\cos\eta\cos\gamma&=\cos\theta'\sin\eta \end{aligned}\right\} \quad . \quad (17)$$

The differential equations of the rotation of a solid body about its centre of gravity, whatever its motion of translation may be, are the following —

$$\left.\begin{aligned} \mathrm{A}\frac{dp}{dt}-(\mathrm{B}-\mathrm{C})\,qr&=\mathrm{P} \\ \mathrm{B}\frac{dq}{dt}-(\mathrm{C}-\mathrm{A})\,rp&=\mathrm{Q} \\ \mathrm{C}\frac{dr}{dt}-(\mathrm{A}-\mathrm{B})\,pq&=\mathrm{R} \end{aligned}\right\} \quad . \quad . \quad . \quad (18)$$

$$\left.\begin{aligned} p&=\sin\phi\sin\theta\frac{d\psi}{dt}-\cos\phi\frac{d\theta}{dt} \\ q&=\cos\phi\sin\theta\frac{d\psi}{dt}+\sin\phi\frac{d\theta}{dt} \\ r&=\frac{d\phi}{dt}-\cos\theta\frac{d\psi}{dt} \end{aligned}\right\} \quad . \quad . \quad . \quad (19)$$

where A, B, C denote the moments of inertia about the principal axes x, y, z, which intersect in the centre of gravity; p, q, r the components of angular velocity of rotation, and P, Q, R the moments of external forces with regard to the same axes.

The disc under consideration being a body of revolution about the axis of $z_{,}$, the moments of inertia about the other two axes are equal, id est

$$\mathrm{B}=\mathrm{A} \quad . \quad . \quad . \quad . \quad . \quad . \quad . \quad . \quad . \quad . \quad (20)$$

With regard to the moments of the forces which act upon it, we must in the first instance remark that gravity must be eliminated, because the force which results from it acts through the centre of moments. The disc receives from the ring motions, the resultant moment of which we will denote by μ. Neglecting the effect of friction, and admitting a perfect symmetry about the axis of the disc, it follows that the forces acting parallel to the axis upon the pivots are reduced to one force, the direction of which passes through the axis of revolution, and has therefore no moment. Likewise neglecting the resistance of the air, there remain forces perpendicular to the surfaces of the pivots, and the common direction of all those forces passes through the generating axis of the whirling body. The axis of the resultant moment μ is therefore situated in the equatorial plane of the disc. Let λ denote the angle which this axis makes with that of $x_{,}$, taken in a direction from $x_{,}$ to $y_{,}$; then we have

$$\left.\begin{array}{l} P=\mu \cos \lambda \\ Q=\mu \sin \lambda \\ R=\mu \cos 90^{\circ}=0 \end{array}\right\} \quad . \quad . \quad . \quad (21)$$

This last value reduces the third equation of (18) by means of (20) to $\frac{dr}{dt}=0$; whence denoting by n a certain constant we have

$$r=n \quad . \quad . \quad . \quad . \quad . \quad . \quad . \quad . \quad . \quad . \quad . \quad . \quad . \quad . \quad . \quad (22)$$

By means of the equations just established, the two first of (18) are transformed into

$$\left.\begin{array}{l} A\frac{dp}{dt}+(C-A)\,n\,q=\mu \cos \lambda \\ A\frac{dq}{dt}-(C-A)\,n\,p=\mu \sin \lambda \end{array}\right\} \quad . \quad (23)$$

The equations of the motion of the ring may be written down at once by accentuating all the terms which enter in the equations (18) and (19), with the exception of the time t. We have therefore from equation (18)

$$\left.\begin{aligned} A'\frac{dp'}{dt}-(B'-C')\,q'r'&=P' \\ B'\frac{dq'}{dt}-(C'-A')\,r'p'&=Q' \\ C'\frac{dr'}{dt}-(A'-B')\,p'q'&=R' \end{aligned}\right\} \quad . \quad . \quad . \quad . \quad (24)$$

and from equation (19) we obtain, because $\theta'=$constant,

$$p'=\sin\phi'\,\sin\theta'\,\frac{d\psi'}{dt}$$

$$q'=\cos\phi'\,\sin\theta'\,\frac{d\psi'}{dt}$$

$$r'=\frac{d\phi'}{dt}-\cos\theta'\,\frac{d\psi'}{dt}\ ;$$

but by means of equations (10) and (1) we have

$$\frac{d\psi}{dt}=-\omega \quad . \quad . \quad . \quad . \quad . \quad . \quad . \quad . \quad . \quad . \quad (25)$$

Moreover from (11) $\phi'=\gamma$; hence by substitution

$$\left.\begin{aligned} p'&=-\omega\sin\theta\sin\gamma \\ q'&=-\omega\sin\theta'\cos\gamma \\ r'&=\frac{d\gamma}{dt}+\omega\cos\theta' \end{aligned}\right\} \quad . \quad . \quad . \quad . \quad (26)$$

therefore by differentiation

$$\left.\begin{aligned} \frac{dp'}{dt}&=-\omega\sin\theta\cos\gamma\,\frac{d\gamma}{dt}, \\ \frac{dq'}{dt}&=+\omega\sin\theta'\sin\gamma\,\frac{d\gamma}{dt}, \\ \frac{dr'}{dt}&=\frac{d^2\gamma}{dt^2}. \end{aligned}\right\} \quad . \quad . \quad . \quad . \quad (27)$$

The moments of the forces which the ring receives are of two kinds. Firstly, the moments of the forces equal and opposite to those which it exerts upon the disc; secondly, the moments of forces which the supports exert upon the axis of the ring itself. The latter may be reduced to a single moment μ', the axis of which is situated in the equatorial plane of the ring, and makes an angle λ' with the axis of $x_{,}'$ taken in the direction from $x_{,}'$ to $y_{,}'$. Therefore

$$\begin{aligned}
P' &= -P\cos(x, x_{,}') - Q\cos(y, x_{,}') + \mu'\cos\lambda' \\
Q' &= -P\cos(x, y_{,}') - Q\cos(y, y_{,}') + \mu'\sin\lambda' \\
R' &= -P\cos(x, z_{,}') - Q\cos(y, z_{,}').
\end{aligned}$$

Taking the values of P and Q from (21) and making further substitutions by means of the equations (2) to (8), we have

$$\begin{aligned}
P' &= -\mu\cos\lambda\sin\alpha - \mu\sin\lambda\cos\alpha + \mu'\cos\lambda' \\
Q' &= +\mu\cos\lambda\cos\alpha\cos\eta - \mu\sin\lambda\sin\alpha\cos\eta + \mu'\sin\lambda' \\
R' &= +\mu\cos\lambda\cos\alpha\sin\eta - \mu\sin\lambda\sin\alpha\sin\eta.
\end{aligned}$$

If we substitute these values, as also the preceding, determined by (26) and (27), into the system of equations (24), we obtain the following expressions —

$$\left.\begin{aligned}
&A'\omega\sin\theta'\cos\gamma\frac{d\gamma}{dt} + (B'-C')\,\omega\sin\theta'\cos\gamma\left(\frac{d\gamma}{dt} + \omega\cos\theta'\right) \\
&\qquad = -\mu\sin(\alpha+\lambda) + \mu'\cos\lambda' \\
&B'\omega\sin\theta'\sin\gamma\frac{d\gamma}{dt} + (C'-A')\,\omega\sin\theta'\sin\gamma\left(\frac{d\gamma}{dt} + \omega\cos\theta'\right) \\
&\qquad = \mu\cos\eta\cos(\alpha+\lambda) + \mu'\sin\lambda' \\
&C'\frac{d^2\gamma}{dt^2} - (A'-B')\,\omega^2\sin^2\theta'\sin\gamma\cos\gamma = \mu\sin\eta\cos(\alpha+\lambda)
\end{aligned}\right\} \quad (28)$$

The first two of these equations may be written

$$\left.\begin{aligned}\mu' \sin \lambda' &= \omega \sin \theta' \sin \gamma \left\{\begin{matrix}(\mathrm{B}'+\mathrm{C}'-\mathrm{A}')\dfrac{d\gamma}{dt} \\ +(\mathrm{C}'-\mathrm{A}')\omega \cos \theta'\end{matrix}\right\} - \mu \cos \eta \cos(\alpha+\lambda) \\ \mu' \cos \lambda' &= \omega \sin \theta' \cos \gamma \left\{\begin{matrix}(\mathrm{B}'-\mathrm{C}'-\mathrm{A}')\dfrac{d\gamma}{dt} \\ +(\mathrm{B}'-\mathrm{C}')\omega \cos \theta'\end{matrix}\right\} + \mu \sin(\alpha+\lambda);\end{aligned}\right\} \quad (29)$$

they serve to determine μ' and λ' when all the other quantities relating to the motion of the disc are given.

In order to obtain expressions for the motion of the system of two bodies, we proceed in the following way. Multiplying the first equation of (23) by $\cos \alpha$, the second by $\sin \alpha$, we have by subtraction

$$\mathrm{A}\left(\cos \alpha \frac{dp}{dt} - \sin \alpha \frac{dq}{dt}\right) + (\mathrm{C}-\mathrm{A})\, n\,(q \cos \alpha + p \sin \alpha) \\ = \mu \cos(\alpha+\lambda) \quad . \quad . \quad . \quad . \quad . \quad . \quad . \quad . \quad . \quad . \quad (30)$$

and by a similar method

$$\mathrm{A}\left(\sin \alpha \frac{dp}{dt} + \cos \alpha \frac{dq}{dt}\right) + (\mathrm{C}-\mathrm{A})n\,(q \sin \alpha - p \cos \alpha) \\ = \mu \sin(\alpha+\lambda) \quad . \quad . \quad . \quad . \quad . \quad . \quad . \quad . \quad . \quad . \quad . \quad (31)$$

These equations are adapted to give μ and $\alpha+\lambda$ in functions of other quantities supposed to be known. The first may also be used to form the principal equation of the problem; for the equation (30), being multiplied by $\sin \eta$ and subtracted from the third equation of (28), gives

$$\left.\begin{aligned}&\mathrm{C}'\frac{d^2\gamma}{dt^2} - (\mathrm{A}'-\mathrm{B}')\,\omega^2 \sin^2 \theta' \sin \gamma \cos \gamma \\ &\quad + \mathrm{A} \sin \eta \left(\sin \alpha \frac{dq}{dt} - \cos \alpha \frac{dp}{dt}\right) \\ &\quad - (\mathrm{C}-\mathrm{A}) \sin \eta\; n\,(q \cos \alpha + p \sin \alpha)\end{aligned}\right\} = 0 \quad (32)$$

This would be the equation from which, if we chose, we had to eliminate all the variable quantities with its derivatives except γ; but for the sake of symmetry we will proceed otherwise.

By means of the values in (19) we find those of the following expressions, which are contained in the equations (30), (31), and (32):—

$$q\cos\alpha+p\sin\alpha=\cos(\phi-\alpha)\sin\theta\frac{d\psi}{dt}+\sin(\phi-\alpha)\frac{d\theta}{dt}$$

$$q\sin\alpha-p\cos\alpha=-\sin(\phi-\alpha)\sin\theta\frac{d\psi}{dt}+\cos(\phi-\alpha)\frac{d\theta}{dt}.$$

But by the equation (9)

$$\phi-\alpha=90^\circ-\epsilon;$$

therefore

$$\left.\begin{aligned} q\cos\alpha+p\sin\alpha &= \sin\epsilon\sin\theta\frac{d\psi}{dt}+\cos\epsilon\frac{d\theta}{dt} \\ q\sin\alpha-p\cos\alpha &= -\cos\epsilon\sin\theta\frac{d\psi}{dt}+\sin\epsilon\frac{d\theta}{dt} \end{aligned}\right\} \quad . \quad (33)$$

By means of the equations (12) and (1) we have the value of $\frac{d\psi}{dt}$ determined thus:—

$$\frac{d\psi}{dt}=\frac{d\beta}{dt}-\omega \quad . \quad . \quad . \quad . \quad . \quad . \quad . \quad . \quad (34)$$

and in order to obtain the value of $\frac{d\beta}{dt}$ we have recourse to the first equation of (14) and to the third of (15), writing them as follows:—

$$\sin\theta\sin\beta=\sin\eta\sin\gamma$$

$$\sin\theta\cos\beta=\frac{\cos\eta}{\sin\theta'}-\cot\theta'\cos\theta.$$

They give, by differentiating them —

$$\cos\beta\sin\theta\frac{d\beta}{dt}+\sin\beta\cos\theta\frac{d\theta}{dt}=\sin\eta\cos\gamma\frac{d\gamma}{dt}$$

$$-\sin\beta\sin\theta\frac{d\beta}{dt}+\cos\beta\cos\theta\frac{d\theta}{dt}=\cot\theta'\sin\theta\frac{d\theta}{dt}$$

multiplying the first by $\cos\beta$, the second by $(-\sin\beta)$, we get by addition —

$$\sin\theta\frac{d\beta}{dt}=\sin\eta\cos\beta\cos\gamma\frac{d\gamma}{dt}-\cos\theta'\sin\beta\sin\theta\frac{d\theta}{dt}$$

but the first equation of (15) gives by differentiation,

$$\sin\theta\frac{d\theta}{dt}=\sin\eta\sin\theta'\sin\gamma\frac{d\gamma}{dt}\quad\dots\dots\quad(35)$$

whence, by substitution —

$$\sin\theta\frac{d\beta}{dt}=\sin\eta\,(\cos\beta\cos\gamma-\sin\beta\sin\gamma\cos\theta')\frac{d\gamma}{dt}$$

or by means of the second equation of (16)

$$\sin\theta\frac{d\beta}{dt}=-\sin\eta\cos\varepsilon\frac{d\gamma}{dt}\quad\dots\dots\quad(36)$$

Substituting the values (34), (35), and (36) into the equations (33), we obtain —

$$q\cos\alpha+p\sin\alpha=-\omega\sin\theta\sin\varepsilon-\frac{\sin\eta\cos\varepsilon}{\sin\theta}(\sin\varepsilon\sin\theta-\sin\theta'\sin\gamma)\frac{d\gamma}{dt}$$

$$q\sin\alpha-p\cos\alpha=\omega\sin\theta\cos\varepsilon+\sin\eta\left(\cos^2\varepsilon+\sin\varepsilon\frac{\sin\theta'\sin\gamma}{\sin\theta}\right)\frac{d\gamma}{dt}.$$

These expressions are reduced by means of the several equations of (14) to —

$$\left.\begin{aligned}q\cos\alpha+p\sin\alpha&=-\omega\sin\theta\sin\varepsilon\\ q\sin\alpha-p\cos\alpha&=+\omega\sin\theta\cos\varepsilon+\sin\eta\frac{d\gamma}{dt}\end{aligned}\right\}\quad(37)$$

Now, differentiating these two equations, we have

$$\left.\begin{aligned}
&\cos\alpha\frac{dq}{dt}+\sin\alpha\frac{dp}{dt}-(q\sin\alpha-p\cos\alpha)\frac{d\alpha}{dt}\\
&\qquad=-\omega\frac{d.\sin\theta\sin\varepsilon}{dt}\\
&\sin\alpha\frac{dq}{dt}-\cos\alpha\frac{dp}{dt}+(q\cos\alpha+p\sin\alpha)\frac{d\alpha}{dt}\\
&\qquad=\omega\frac{d.\sin\theta\cos\varepsilon}{dt}+\sin\eta\frac{d^2\gamma}{dt^2}
\end{aligned}\right\}\qquad(38)$$

The second equations of (14) and (15) give respectively

$$\sin\theta\sin\varepsilon=\sin\theta'\sin\gamma$$

$$\sin\theta\cos\varepsilon=\frac{\cos\theta'}{\sin\eta}-\cot\eta\cos\theta$$

whence, by differentiation,

$$\left.\begin{aligned}
\frac{d.\sin\theta\sin\varepsilon}{dt}&=\cos\varepsilon\sin\theta\frac{d\varepsilon}{dt}+\sin\varepsilon\cos\theta\frac{d\theta}{dt}\\
&=\sin\theta'\cos\gamma\frac{d\gamma}{dt}\\
\frac{d.\sin\theta\cos\varepsilon}{dt}&=-\sin\varepsilon\sin\theta\frac{d\varepsilon}{dt}+\cos\varepsilon\cos\theta\frac{d\theta}{dt}\\
&=\cot\eta\sin\theta\frac{d\theta}{dt}
\end{aligned}\right\}\qquad(39)$$

Multiplying the first of these equations by $\cos\varepsilon$, the second by $(-\sin\varepsilon)$, we have by addition

$$\sin\theta\frac{d\varepsilon}{dt}=\sin\theta'\cos\gamma\cos\varepsilon\frac{d\gamma}{dt}-\cot\eta\sin\varepsilon\sin\theta\frac{d\theta}{dt}$$

and, substituting the value of (35)

$$\sin\theta\frac{d\varepsilon}{dt}=\sin\theta'(\cos\gamma\cos\varepsilon-\sin\gamma\sin\varepsilon\cos\eta)\frac{d\gamma}{dt}$$

which by means of the first equation of (16) is reduced to

$$\sin\theta\frac{d\varepsilon}{dt}=-\sin\theta'\cos\beta\frac{d\gamma}{dt}. \quad . \quad . \quad . \quad (40)$$

We must now find an expression for the differential co-efficient $\frac{d\alpha}{dt}$ which is contained in the equations (38). The equation (9) gives

$$d\phi-d\alpha=-d\varepsilon;$$

whence

$$\frac{d\alpha}{dt}=\frac{d\phi}{dt}+\frac{d\varepsilon}{dt};$$

and by means of equation (22) and the third of (19) we have

$$\frac{d\alpha}{dt}=n+\cos\theta\frac{d\psi}{dt}+\frac{d\varepsilon}{dt};$$

hence, by substitution,

$$\frac{d\alpha}{dt}=n-\omega\cos\theta-\frac{1}{\sin\theta}(\sin\eta\cos\theta\cos\varepsilon+\sin\theta'\cos\beta)\frac{d\gamma}{dt},$$

or by means of the first equations of (15) and (17)

$$\left.\begin{aligned}\frac{d\alpha}{dt}&=n-\omega\cos\theta-\cos\eta\frac{d\gamma}{dt}\\ &=n-\omega(\cos\eta\cos\theta'+\sin\eta\sin\theta'\cos\gamma)-\cos\eta\frac{d\gamma}{dt}\end{aligned}\right\}(41)$$

With regard to the constant quantity n, we must remark that it is not directly given us by observation, but the preceding equation will furnish us with the means to obtain it. Let us suppose that the ring is at rest with regard to the surface of the earth, making with the horary plane an angle γ_0, and that we communicate to the disc an angular velocity relatively to the plane of the ring expressed by w; then this quantity will be the initial value of $\frac{d\alpha}{dt}$, and the equation (41) furnishes the following expression for n in functions of w and γ_0:—

$$n = w + \omega(\cos\eta\cos\theta' + \sin\eta\sin\theta'\cos\gamma_0) = w + \omega\cos\theta_0 \quad (42)$$

This equation shows that the velocity of rotation about the axis of the disc is equal to the relative velocity of the disc, the ring being at rest, increased by the component of the velocity of the earth about the axis of the disc, the ring likewise considered fixed.

Substituting the values (39), (35), and (41) into the equations (38) we have

$$\cos\alpha\frac{dq}{dt} + \sin\alpha\frac{dp}{dt} = \left(n - \omega\cos\theta - \cos\eta\frac{d\gamma}{dt}\right)(q\sin\alpha - p\cos\alpha)$$
$$-\omega\sin\theta'\cos\gamma\frac{d\gamma}{dt}$$

$$\sin\alpha\frac{dq}{dt} - \cos\alpha\frac{dp}{dt} = -\left(n - \omega\cos\theta - \cos\eta\frac{d\gamma}{dt}\right)(q\cos\alpha + p\sin\alpha)$$
$$+\omega\sin\theta'\cos\eta\sin\gamma\frac{d\gamma}{dt} + \sin\eta\frac{d^2\gamma}{dt^2}.$$

By means of these values the equations (31) and (30) are transformed into

$$\mu\sin(\alpha+\lambda) = \left[Cn - A\left(\omega\cos\theta + \cos\eta\frac{d\gamma}{dt}\right)\right](q\sin\alpha - p\cos\alpha)$$
$$-A\omega\sin\theta'\cos\gamma\frac{d\gamma}{dt}$$

$$\mu\cos(\alpha+\lambda) = \left[Cn - A\left(\omega\cos\theta + \cos\eta\frac{d\gamma}{dt}\right)\right](q\cos\alpha + p\sin\alpha)$$
$$-A\omega\sin\theta'\cos\eta\sin\gamma\frac{d\gamma}{dt} - A\sin\eta\frac{d^2\gamma}{dt^2}.$$

Substituting the values (37) we have

$$\mu \sin(\alpha+\lambda) = \left[Cn - A\left(\omega \cos\theta + \cos\eta \frac{d\gamma}{dt}\right)\right]\left(\omega \sin\theta \cos\epsilon + \sin\eta \frac{d\gamma}{dt}\right)$$

$$- A\omega \sin\theta' \cos\gamma \frac{d\gamma}{dt}.$$

$$\mu \cos(\alpha+\lambda) = -\left[Cn - A\left(\omega \cos\theta + \cos\eta \frac{d\gamma}{dt}\right)\right] \omega \sin\theta \sin\epsilon$$

$$- A\omega \sin\theta' \cos\eta \sin\gamma \frac{d\gamma}{dt} - A \sin\eta \frac{d^2\gamma}{dt^2};$$

and, developing and arranging —

$$\mu \sin(\alpha+\lambda) = (Cn - A\omega \cos\theta)\, \omega \sin\theta \cos\varepsilon$$
$$+ [Cn \sin\eta - A\omega(\sin\eta \cos\theta + \sin\theta' \cos\gamma$$
$$+ \cos\eta \sin\theta \cos\varepsilon)] \frac{d\gamma}{dt} - A \sin\eta \cos\eta \frac{d\gamma^2}{dt^2}$$

$$\mu \cos(\alpha+\lambda) = -(Cn - A\omega \cos\theta)\, \omega \sin\theta \sin\varepsilon$$
$$+ A\omega \cos\eta\, (\sin\theta \sin\varepsilon - \sin\theta' \sin\gamma) \frac{d\gamma}{dt}$$
$$- A \sin\eta \frac{d^2\gamma}{dt^2}.$$

By means of the second equations of (14) and (17) these expressions are reduced into

$$\left.\begin{aligned}
\mu \sin(\alpha+\lambda) &= (Cn - A\omega \cos\theta)\, \omega \sin\theta \cos\varepsilon \\
&\quad + (Cn - 2\, A\omega \cos\theta) \sin\eta \frac{d\gamma}{dt} \\
&\quad - A \sin\eta \cos\eta \frac{d\gamma^2}{dt^2} \\
\mu \cos(\alpha+\lambda) &= -(Cn - A\omega \cos\theta)\, \omega \sin\theta' \sin\gamma \\
&\quad - A \sin\eta \frac{d^2\gamma}{dt^2}.
\end{aligned}\right\} \quad (43)$$

If we put for $\cos\theta$ its value from (15) and for $\sin\theta \cos\varepsilon$ that derived from the third equation of (17), the formulæ (43) will give μ and $\alpha+\lambda$ in functions of γ and its two first derivatives.

In order to form the final equation, we eliminate the quantity $\mu \cos(\alpha+\lambda)$ between the third equation of (28) and the second of (43), thus we obtain

$$(\mathrm{A} \sin^2 \eta+\mathrm{C}')\frac{d^2\gamma}{dt^2}+(\mathrm{C}n-\mathrm{A}\omega \cos\theta)\,\omega \sin\theta' \sin\eta \sin\gamma$$
$$-(\mathrm{A}'-\mathrm{B}')\,\omega^2 \sin^2\theta' \sin\gamma \cos\gamma=0;$$

then replacing $\cos\theta$ by its value (15), and transposing, we have

$$(\mathrm{A} \sin^2\eta+\mathrm{C}')\frac{d^2\gamma}{dt^2}=-(\mathrm{C}n-\mathrm{A}\cos\eta.\omega\cos\theta')\,\omega\sin\theta' \sin\eta\sin\gamma \quad (44)$$
$$+(\mathrm{A}\sin^2\eta+\mathrm{A}'-\mathrm{B}')\,\omega^2 \sin^2\theta' \sin\gamma\cos\gamma.$$

For brevity's sake we put

$$\frac{g}{a}=\frac{\mathrm{C}n-\mathrm{A}\cos\eta\,.\,\omega\cos\theta'}{\mathrm{A}\sin^2\eta+\mathrm{C}'}\,\omega\sin\theta'\sin\eta$$
$$2\delta\frac{g}{a}=\frac{\mathrm{A}\sin^2\eta+\mathrm{A}'-\mathrm{B}'}{\mathrm{A}\sin^2\eta+\mathrm{C}'}\,\omega\sin^2\theta'$$

Where g denotes the accelerating force of gravity, and δ some abstract number, which will always be very small, the preceding formula is then written

$$\frac{d^2\gamma}{dt^2}=\frac{g}{a}\left(-\sin\gamma+2\delta\sin\gamma\cos\gamma\right) \quad . \quad . \quad . \quad . \quad (45)$$

and we have at the same time

$$\left.\begin{aligned} a&=\frac{g}{\omega\sin\theta'\sin\eta}\cdot\frac{\mathrm{A}\sin^2\eta+\mathrm{C}'}{\mathrm{C}n-\mathrm{A}\cos\eta\,.\,\omega\cos\theta'},\\ \delta&=\tfrac{1}{2}\,\frac{\omega\sin\theta'}{\sin\eta}\cdot\frac{\mathrm{A}\sin^2\eta+\mathrm{A}'-\mathrm{B}'}{\mathrm{C}n-\mathrm{A}\cos\eta\,.\,\omega\cos\theta'}.\end{aligned}\right\} \quad (46)$$

If $\pm\gamma_0$ be the value of γ, for which $\frac{d\gamma}{dt}$ becomes zero, we get by multiplying the equation (45) by $2\,d\gamma$, and then integrating—

$$\frac{d\gamma^2}{dt^2}=\frac{2g}{a}\left[\cos\gamma-\cos\gamma_0-\delta\left(\cos^2\gamma-\cos^2\gamma_0\right)\right]$$

or

$$\frac{d\gamma}{dt}=\pm\sqrt{\frac{2g}{a}(\cos\gamma-\cos\gamma_0)[1-\delta(\cos\gamma+\cos\gamma_0)]} \quad (47)$$

The case where γ becomes imaginary, for a value $\frac{d\gamma}{dt}=0$ is excluded from our considerations, therefore γ will always be a real angle, and because δ is very small, the factor $1-\delta\ (\cos\gamma+\cos\gamma_0)$ will always be positive; hence the only condition for $\frac{d\gamma}{dt}$ to be a real quantity is that the factor $(\cos\gamma-\cos\gamma_0)$ and the constant (a) have the same sign. Now, as the angles θ' and η are supposed to lie between zero and 180°, and the term $A\cos\eta\,.\,\omega\cos\theta'$ is very small compared with Cn, the sign of (a) will be identical with that of (n). When therefore the component n of the rotation about the axis of the disc is positive, we have

$$\cos\gamma > \cos\gamma_0,$$

and the values of γ follow in the order

$$\gamma_0,\ 0,\ -\gamma_0,\ 0,\ +\gamma_0,\ 0,\ \ldots\ \&c.,$$

but if the motion takes place in an opposite direction, that is, n being negative, we have

$$\cos\gamma < \cos\gamma_0,$$

and the values of γ succeed each other as follows :—

$$\gamma_0,\ \pi,\ 2\pi-\gamma_0,\ \gamma_0,\ \pi,\ 2\pi-\gamma_0,\ \ldots\ \&c.,$$

The absolute value of $\frac{d\gamma}{dt}$ is the same for two equal values of γ with contrary signs; hence the motion of the plane of the ring is an oscillation about the horary plane, and the angular velocities are equal in positions symmetrical with regard to this plane. Neglecting the term affected with δ in equation (47), we find that the expression coincides with the differential equation for the motion of the pendulum of length a. The oscillations of the ring follow in this case the same laws as the rotation of the pendulum about the vertical. We will now consider separately the two cases of n being positive or negative.

Let n be positive and

$$\xi = \cos\gamma,\ \xi_0 = \cos\gamma_0 \quad . \quad . \quad . \quad . \quad . \quad . \quad . \quad (48)$$

whence

$$\sin\gamma\, d\gamma = -d\xi,$$

and

$$d\gamma = \pm \frac{1}{\sqrt{1-\xi^2}}\, d\xi.$$

The equation (47) gives therefore

$$\sqrt{\frac{2g}{a}}\, dt = \pm \frac{d\xi}{\sqrt{(1-\xi^2)(\xi-\xi_0)\left[1-\delta(\xi+\xi_0)\right]}} \quad (49)$$

Let us denote by N the value of the radical on the right side of this equation, and suppose

$$(\xi-\xi_0)\left[1-\delta(\xi+\xi_0)\right] = (1-\xi^2)\, z^2 \quad . \quad . \quad . \quad (50)$$

so that we have

$$N = (1-\xi^2)\, z \quad . \quad . \quad . \quad . \quad . \quad . \quad (51)$$

then the equation to be integrated becomes

$$\sqrt{\frac{2g}{a}}\, dt = \frac{d\xi}{N} \quad . \quad . \quad . \quad . \quad . \quad (52)$$

From the equation (50) we derive

$$\xi\,(z^2-\delta)=-\tfrac{1}{2}+Z\,. \quad . \quad . \quad . \quad . \quad (53)$$

where

$$\begin{aligned} Z^2 &= \tfrac{1}{4}\,[\,1-4\,\delta\,\xi_0\,(1-\delta\,\xi_0)]+[\,\xi_0\,(1-\delta\,\xi_0)-\delta]z^2+z^4 \\ &= \tfrac{1}{4}\,(1-2\,\delta\,\xi_0)^2+[\xi_0-\delta\,(1+\xi^2{}_0)]\;z^2+z^4. \end{aligned}$$

Differentiating the equation (50) we obtain

$$[1+2\,\xi\,(z^2-\delta)]\,d\,\xi=2\,(1-\xi^2)\,z\,d\,z\,;$$

whence by means of the equations (51) and (53)

$$\frac{d\,\xi}{\mathrm{N}}=\frac{d\,z}{Z};$$

and therefore

$$\sqrt{\frac{2\,g}{a}}\,d\,t=\frac{d\,z}{Z}\,. \quad . \quad . \quad . \quad . \quad (54)$$

Observing that if neglecting the terms affected with δ the value of Z^2 is greater than $\frac{1}{4}\,\xi_0{}^2+\xi_0\,z^2+z^4=(\frac{1}{2}\,\xi_0+z^2)^2$, a quantity essentially positive, we may assume

$$Z^2=\upsilon^2+2\,\upsilon\,\cos\,2\,\kappa\,.\,z^2+z^4\,. \quad . \quad . \quad (54\text{ bis})$$

In order that this value may agree with the preceding, we must put

$$\begin{gathered} \upsilon^2=\tfrac{1}{4}\,(1-2\,\delta\,\xi_0)^2 \\ 2\,\upsilon\,\cos\,2\,\kappa=\xi_0-\delta\,(1+\xi_0{}^2) \end{gathered}$$

whence

$$2\,\upsilon=1-2\,\delta\,\xi_0$$

$$\cos\,2\,\kappa=\frac{\xi_0-\delta\,(1+\xi_0{}^2)}{1-2\,\delta\,\xi_0}\,. \quad . \quad . \quad . \quad (55)$$

from which we derive

$$\sin^2\kappa=\frac{1-\xi_0+\delta\,(1-2\,\xi_0+\xi_0{}^2)}{2\,(1-2\,\delta\xi_0)}=\frac{1-\xi_0}{2}\;\frac{1+\delta\,(1-\xi_0)}{1-2\delta\,\xi_0}$$

or because

$$\xi_0=\cos\,\gamma_0$$

$$\sin^2 \kappa = \sin^2 \tfrac{1}{2}\gamma_0 \frac{1+2\delta \sin^2 \frac{1}{2}\gamma_0}{1-2\delta \cos \gamma_0},$$

and

$$\sin \kappa = \pm \sin \tfrac{1}{2}\gamma_0 \sqrt{\frac{1+2\delta \sin^2 \frac{1}{2}\gamma_0}{1-2\delta \cos \gamma_0}} \quad . \quad . \quad (56)$$

Taking

$$z = \sqrt{v} \text{ tang } \tfrac{1}{2}\phi \quad . \quad . \quad . \quad . \quad . \quad . \quad (57)$$

we have by differentiation

$$dz = \tfrac{1}{2} \sqrt{v} \frac{d\phi}{\cos^2 \frac{1}{2}\phi} \quad . \quad . \quad . \quad . \quad . \quad . \quad (58)$$

the value of Z^2 (54 bis) is then written

$$Z^2 = v^2 (1 + 2 \cos 2\kappa \text{ tang}^2 \tfrac{1}{2}\phi + \tan^4 \tfrac{1}{2}\phi)$$

$$= \frac{v^2}{\cos^4 \frac{1}{2}\phi} \left[\cos^4 \tfrac{1}{2}\phi + 2 \ (1 - 2 \sin^2 \kappa) \sin^2 \tfrac{1}{2}\phi \cos^2 \tfrac{1}{2}\phi + \sin^4 \tfrac{1}{2}\phi \right]$$

$$= \frac{v^2}{\cos^4 \frac{1}{2}\phi} \left[(\cos^2 \tfrac{1}{2}\phi + \sin^2 \tfrac{1}{2}\phi)^2 - 4 \sin^2 \kappa \sin^2 \tfrac{1}{2}\phi \cos^2 \tfrac{1}{2}\phi \right]$$

and if we write $c^2 = \sin^2 \kappa$, we have simply

$$Z = \pm \frac{v}{\cos^2 \frac{1}{2}\phi} \sqrt{1 - c^2 \sin^2 \phi} \quad . \quad (59)$$

whence by means of equation (58)

$$\frac{dz}{Z} = \pm \frac{1}{2\sqrt{v}} \frac{d\phi}{\sqrt{1 - c^2 \sin^2 \phi}} \quad . \quad . \quad . \quad . \quad (60)$$

and by means of this value and equation (54)

$$dt = \frac{1}{2\sqrt{1 - 2\delta \cos \gamma_0}} \sqrt{\frac{a}{g}} \cdot \frac{d\phi}{\sqrt{1 - c^2 \sin^2 \phi}} \quad . \quad (61)$$

We have here suppressed the double sign, because the element of time, dt, is essentially positive, and we are going to determine the angle ϕ, so that $d\phi$ may also be positive. For this purpose we derive from the equations (50), (55), and (57)

$$\tan \tfrac{1}{2}\phi = \pm \sqrt{2\frac{(\xi-\xi_0)\left[1-\delta(\xi+\xi_0)\right]}{(1-\xi^2)(1-2\delta\xi_0)}}$$

or by means of equations (48)

$$\tan \tfrac{1}{2}\phi = \pm \frac{1}{\sin\gamma}\sqrt{\frac{2(\cos\gamma-\cos\gamma_0)\left[1-\delta(\cos\gamma+\cos\gamma_0)\right]}{1-2\delta\cos\gamma_0}}. \quad (62)$$

This value may also be written:

$$\tan \tfrac{1}{2}\phi = \pm \frac{2}{\sin\gamma}\sqrt{\frac{\sin\frac{1}{2}(\gamma_0-\gamma)\sin\frac{1}{2}(\gamma_0+\gamma)}{1-2\delta\cos\gamma_0}\left[1-2\delta\cos\tfrac{1}{2}(\gamma_0+\gamma)\cos\tfrac{1}{2}(\gamma_0-\gamma)\right]}$$

We have seen that the values of γ succeed each other in the order

$$+\gamma_0,\ 0,\ -\gamma_0,\ 0,\ +\gamma_0,\ 0,\ -\gamma_0 \ldots \text{ etc.}$$

Taking $\sin\gamma_0$ to be positive, let us write down the corresponding values of $\tan\frac{1}{2}\phi$, with the signs which they take in the interval of two consecutive values of γ, and we shall have the series:

$$0,\ +\infty,\ -0,\ +\infty,\ -0,\ +,\infty,\ -0, \ldots \text{ etc.}$$

the corresponding values of $\frac{1}{2}\phi$ are

$$0,\ \frac{\pi}{2},\ \pi,\ 3\frac{\pi}{2}\ 2\pi,\ 5\frac{\pi}{2},\ 3\pi, \ldots \text{ etc.}$$

and those of ϕ:

$$0,\ \pi,\ 2\pi,\ 3\pi,\ 4\pi,\ 5\pi,\ 6\pi, \ldots \text{ etc.}$$

The passage from γ_0 to $-\gamma_0$, and conversely, corresponds to a variation of ϕ, which is equal to 2π. If we take $\sin\gamma_0$ to be negative, we obtain similarly an ascending and continuous series of values of ϕ.

Integrating the equation (61), we count the time from

that moment when $\gamma=\gamma_0$, and take simultaneously ϕ equal to zero; thus we have

$$t=\frac{1}{2\sqrt{1-2\,\delta\cos\gamma_0}}\cdot\sqrt{\frac{a}{g}}\cdot \mathrm{F}\,(c,\phi). \quad . \quad . \quad (63)$$

In order to obtain the duration of the simple oscillation between the limits $+\gamma_0$ and $-\gamma_0$, we must make $\phi=2\,\pi$; but since

$$\mathrm{F}\,(c,\,2\pi)=4\mathrm{F}\left(c,\frac{\pi}{2}\right),$$

we have, denoting by T the duration of oscillation,

$$\mathrm{T}=\frac{2}{\sqrt{1-2\delta\cos\gamma_0}}\sqrt{\frac{a}{g}}\,\mathrm{F}\left(c,\frac{\pi}{2}\right), \quad . \quad . \quad . \quad (64)$$

When we neglect δ, and suppose γ_0 infinitely small, we are led to the common formula of the pendulum:

$$\mathrm{T}=\pi\sqrt{\frac{a}{g}}$$

Let n be negative.

Sin η and sin θ' being positive, and ω very small compared with n; the value of a in the first equation of (46) becomes negative. In this case the expression ($\cos\gamma-\cos\gamma_0$) must also be negative, so that the equation (49) is changed into:

$$\sqrt{\frac{2g}{-a}}\,dt=\pm\frac{d\xi}{\sqrt{(1-\xi^2)\,(\xi_0-\xi)\,[1-\delta\,(\xi_0+\xi)]}} \quad (66)$$

By comparing the right side of this equation with that of equation (49), we see that we pass from one to the other by changing the signs of the quantities ξ, ξ_0, and δ, whence we must change γ into $(180-\gamma)$ and γ_0 into $(180-\gamma_0)$, so that the formulas (56) and (62) are transformed into

$$\left.\begin{aligned} c=\sin\kappa &= \pm\cos\tfrac{1}{2}\gamma_0\sqrt{\frac{1-2\delta\cos^2\frac{1}{2}\gamma_0}{1-2\delta\cos\gamma_0}} \\ \text{tang}\,\tfrac{1}{2}\phi &= \pm\frac{1}{\sin\gamma}\sqrt{\frac{2(\cos\gamma_0-\cos\gamma)[1-\delta(\cos\gamma_0+\cos\gamma)]}{1-2\delta\cos\gamma_0}} \end{aligned}\right\}(67)$$

The values of γ, supposing, for instance, γ_0 to be positive, follow in the order

$$\gamma_0,\ \pi,\ 2\pi-\gamma,\ \pi,\ \gamma_0,\ \pi,\ 2\pi-\gamma_0,\ \pi,\ \gamma_0,\ \ldots\ldots \text{ etc.}$$

The corresponding series of the values of $\tan\frac{1}{2}\phi$ is

$$0,\ +\infty,-0,\ +\infty,\ 0,\ +\infty\ \ldots\ldots \text{ etc.}$$

and the values of ϕ corresponding to it are

$$0\quad \pi\quad 2\pi\quad 3\pi\quad 4\pi\quad 5\pi\ \ldots\ldots\ldots \text{ etc.}$$

so that the general value of t (63) is transformed into:

$$t=\frac{1}{2\sqrt{1-2\delta\cos\gamma_0}}\sqrt{\frac{-a}{g}}\,.\,\mathrm{F}(c,\phi)\quad . \quad (68)$$

The duration of an oscillation corresponding to the limits γ_0 and $(2\pi-\gamma_0)$ of γ, or to the limits zero and 2π of ϕ, is expressed by

$$\mathrm{T}=\frac{2}{\sqrt{1-2\delta\cos\gamma_0}}\sqrt{\frac{-a}{g}}\,.\,\mathrm{F}\left(c,\frac{\pi}{2}\right)\quad . \quad (69)$$

Neglecting δ, and supposing γ_0 infinitely small, the quantity c, becomes equal to unity, and the function $\mathrm{F}\left(1,\frac{\pi}{2}\right)$ is infinite. The plane of the ring therefore being placed without velocity in coincidence with the horary plane, rests in this position, when n is negative; but the equilibrium is unstable, for the least change of the position, or, in other words, the smallest value of γ_0 makes the value of T finite.

In order to compute α, let us add the equations (41) and (42), which gives—

$$\frac{d\alpha}{dt}=w-\omega \sin \theta' \sin \eta (\cos\gamma-\cos\gamma_0)-\cos\eta \frac{d\gamma}{dt}$$

whence we derive

$$\alpha=w\,t-\gamma \cos\eta-\omega \sin\theta' \sin\eta \int(\cos\gamma-\cos\gamma_0)\,dt.$$

In order to effect the integration indicated in the last term, let us first consider the case where (n) and (a) are both positive. We derive from the equation (62) by means of (55)

$$\cos\gamma-\cos\gamma_0=\upsilon \sin^2\gamma \tan^2 \tfrac{1}{2}\phi+\delta\,(\cos^2\gamma-\cos^2\gamma_0) \qquad (71)$$

and since $\xi=\cos\gamma$, the equations (53), (57) and (59) give:

$$\cos\gamma=\frac{-\frac{1}{2}+Z}{\upsilon \tan^2 \frac{1}{2}\phi-\delta}=\frac{-\frac{1}{2}\cos^2\frac{1}{2}\phi \pm \upsilon\sqrt{1-c^2\sin^2\phi}}{\upsilon \sin^2\frac{1}{2}\phi-\delta\cos^2\frac{1}{2}\phi}$$

Since we must have $\cos\gamma=\cos\gamma_0$ for $\phi=0$, the radical sign must be taken positive, for only then will the preceding equation, by substituting the value of υ, give the result

$$\cos\gamma=\frac{-\frac{1}{2}+\frac{1}{2}(1-2\,\delta\cos\gamma_0)}{-\delta}=\cos\gamma_0$$

We have, therefore, in general

$$\cos\gamma=\frac{\upsilon\sqrt{1-c^2\sin^2\phi}-\frac{1}{2}\cos^2\frac{1}{2}\phi}{\upsilon\sin^2\frac{1}{2}\phi-\delta\cos^2\frac{1}{2}\phi}$$

The term to be integrated being affected with a factor (ω) which is extremely small compared with (w), we may neglect δ in the differential expression, which reduces the value of (υ) to $\frac{1}{2}$ and the equation (71) gives,

$$\int(\cos\gamma-\cos\gamma_0)\,dt=\tfrac{1}{2}\int\sin^2\gamma\tan^2\tfrac{1}{2}\phi\,dt.$$

and the value of $\cos\gamma$ itself becomes

$$\cos\gamma=\frac{\sqrt{1-c^2\sin^2\phi}-\cos^2\frac{1}{2}\phi}{\sin^2\frac{1}{2}\phi},$$

from which we derive

$$\sin^2\gamma=\frac{\sin^4\frac{1}{2}\phi-\cos^4\frac{1}{2}\phi-(1-c^2\sin^2\phi)+2\cos^2\frac{1}{2}\phi\sqrt{1-c^2\sin^2\phi}}{\sin^4\frac{1}{2}\phi}$$

$$=\frac{(-\cos^2\frac{1}{2}\phi-\sin^2\frac{1}{2}\phi)-1+4c^2\sin^2\frac{1}{2}\phi\cos^2\frac{1}{2}\phi+2\cos^2\frac{1}{2}\phi\sqrt{1-c^2\sin^2\phi}}{\sin^4\frac{1}{2}\phi}$$

$$=2\frac{\cos^2\frac{1}{2}\phi}{\sin^4\frac{1}{2}\phi}\left(2c^2\sin^2\tfrac{1}{2}\phi-1+\sqrt{1-c^2\sin^2\phi}\right),$$

and the integral in question becomes

$$\int(\cos\gamma-\cos\gamma_0)\,dt=2c^2t-\int\frac{1}{\sin^2\frac{1}{2}\phi}\,dt+\int\frac{\sqrt{1-c^2\sin^2\phi}}{\sin^2\frac{1}{2}\phi}\,dt.$$

Substituting the value of dt from (61), neglecting the term in δ, and taking for brevity's sake

$$\Delta=\sqrt{1-c^2\sin^2\phi},$$

we have

$$\int(\cos\gamma-\cos\gamma_0)\,dt=2c^2t-\sqrt{\frac{a}{g}}\left[\int\frac{1}{2\sin^2\frac{1}{2}\phi}\cdot\frac{d\phi}{\Delta}-\int\frac{d\phi}{2\sin^2\frac{1}{2}\phi}.\right.$$

The first of the last two integrals may be written

$$\int\frac{1}{2\sin^2\frac{1}{2}\phi}\frac{d\phi}{\Delta}=\int\frac{1+\cos\phi}{\sin^2\phi}\cdot\frac{d\phi}{\Delta}$$

$$=\int\frac{1}{\sin^2\phi}\frac{d\phi}{\Delta}+\int\frac{\cos\phi}{\sin^2\phi}\frac{d\phi}{\Delta}$$

In order to effect the integrations indicated on the right side of the equation, let us differentiate the expression

$$\Delta\cot\phi=\sqrt{1-c^2\sin^2\phi}\,.\,\cot\phi$$

thus we have

$$d.\Delta.\cot\phi = -\sqrt{1-c^2\sin^2\phi}\cdot\frac{d\phi}{\sin^2\phi} - \frac{c^2\cos^2\phi}{\sqrt{1-c^2\sin^2\phi}}\,d\phi,$$

$$= -\frac{1-c^2\sin^2\phi + c^2\sin^2\phi\cos^2\phi}{\sin^2\phi\sqrt{1-c^2\sin^2\phi}}.\,d\phi$$

$$= -\frac{1-c^2\sin^4\phi}{\sin^2\phi\,\sqrt{1-c^2\sin^2\phi}}.\,d\phi$$

$$= -\frac{1}{\sin^2\phi}\frac{d\phi}{\Delta} + c^2\frac{\sin^2\phi}{\Delta}\,d\phi,$$

whence we derive

$$\int\frac{1}{\sin^2\phi}\cdot\frac{d\phi}{\Delta} = c^2\int\frac{\sin^2\phi}{\Delta}\,d\phi - \Delta\cot\phi.$$

Let us now take

$$x = c\sin\phi$$

whence we have by differentiation

$$dx = c\,.\,\cos\phi\,.\,d\phi,$$

or,

$$\cos\phi\,.\,d\phi = \frac{dx}{c}$$

an

$$\frac{\cos\phi}{\sin^2\phi}\,\frac{d\phi}{\Delta} = \frac{c}{x^2}\,\frac{dx}{\sqrt{1-x^2}}$$

from which we conclude that

$$\int\frac{\cos\phi}{\sin^2\phi}\cdot\frac{d\phi}{\Delta} = -\frac{c}{x}\sqrt{1-x^2}$$

$$= -\frac{\Delta}{\sin\phi}.$$

We have therefore

$$\int\frac{1}{2\sin^2\frac{1}{2}\phi}\cdot\frac{d\phi}{\Delta} = \int\frac{1}{\sin^2\phi}\,\frac{d\phi}{\Delta} + \int\frac{\cos\phi}{\sin^2\phi}\,\frac{d\phi}{\Delta}$$

$$= c^2\int\frac{\sin^2\phi}{\Delta}\,d\phi - \Delta\cot\phi - \frac{\Delta}{\sin\phi}$$

$$= c^2\int\frac{\sin^2\phi}{\Delta}\,d\phi - \frac{\Delta}{\sin\phi}\,(1+\cos\phi).$$

We have further by direct integration

$$-\int \frac{d\phi}{2\sin^2\frac{1}{2}\phi} = \frac{\cos\frac{1}{2}\phi}{\sin\frac{1}{2}\phi} = \frac{1+\cos\phi}{\sin\phi}.$$

Hence by addition:

$$\int \frac{1}{2\sin^2\frac{1}{2}\phi}\frac{d\phi}{\Delta} - \int \frac{d\phi}{2\sin^2\frac{1}{2}\phi} = c^2\int \frac{\sin^2\phi}{\Delta}\,d\phi + \frac{1+\cos\phi}{\sin\phi}(1-\Delta)$$

In order to avoid the indefinite expression which presents itself in the case of $\sin\phi$ being equal to zero, we multiply the numerator and denominator of the last term by $(1+\Delta)$, thus:

$$\frac{1+\cos\phi}{\sin\phi}(1-\Delta) = \frac{1+\cos\phi}{\sin\phi}\cdot\frac{c^2\sin^2\phi}{1+\Delta} = 2c^2\,\frac{\sin\phi\,.\cos^2\frac{1}{2}\phi}{1+\Delta}$$

so that we have

$$\int(\cos\gamma - \cos\gamma_0)\,dt = 2c^2t - c^2\sqrt{\frac{a}{g}}\left[\int\frac{\sin^2\phi}{\Delta}\,d\phi + \frac{2\sin\phi\cos^2\frac{1}{2}\phi}{1+\Delta}\right]$$

Substituting this value into the equation (70), and supposing that for $t=0$ we have $\alpha=0$, $\gamma=\gamma_0$, $\phi=0$, we obtain as the final result:

$$\left.\begin{aligned}\alpha &= (w - 2\,\omega\sin\theta'\sin\eta\, c^2)\,t - (\gamma-\gamma_0)\cos\eta \\ &\quad + \omega\sin\theta'\sin\eta\, c^2\sqrt{\frac{a}{g}}\left[\int_0^\phi \frac{\sin^2\phi}{\Delta}\,d\phi + \frac{2\sin\phi\cos^2\frac{1}{2}\phi}{1+\Delta}\right]\end{aligned}\right\}\quad(72)$$

In the case of (a) or (n) being negative, it suffices to change the signs of (a) and (ω), and computing c and ϕ by means of equation (67).

The formula (72) shows that the angle α undergoes, besides periodical variations, others which are proportional to the lapse of time, their magnitude depending on the amplitude of oscillations.

CHAPTER IV.

NEW THEORY OF ROTATION.

THE theory of rotation of a rigid system is strictly deduced from the elementary laws of motion, but the complexity of the motion of the particles of a body freely rotating about a point, renders the subject so intricate that it has never been thoroughly understood but by the most expert mathematicians. Many who had mastered the lunar theory came to erroneous conclusions on this subject. Even Newton has chosen to deduce the disturbances of the earth's axis from his theory of the motion of the nodes of a free orbit, rather than attack the problem of the rotation of a solid body.

The method by which Poinsot has rendered the theory more manageable is by a judicious introduction of ideas, chiefly of a geometrical character; and further progress is only to be made by proceeding from one distinct idea to another, instead of trusting to symbols and equations.

It has been shown in the preceding chapters that the earth is endowed with rotation on an axis which has itself a conical motion. On account of the complexity of these two movements, they have been illustrated by a description of Burr's apparatus for imitating the precession of the equinoxes (page 145), but in order to obtain a full understanding of them a perusal of Poinsot's new Theory of Rotation will be the most conducive.

It may, therefore, be of service to many readers to give at the conclusion of this work an outline of that theory, which is most remarkable, both for the lucidity of its reasoning and the importance of its results.

A simple rotatory motion about a fixed axis is easily conceived; all the points of the body describe circles about that axis with an angular velocity, which is the same for all the points at any moment.

But any number of such rotatory motions may be given to the system at once.

If the axis of these rotations meet in a point, the resultant of all the rotatory motions will be a simple rotation about an axis, determined by a theorem analogous to that of the parallelogram of forces (see page 91).

If the axes are parallel to each other, the resultant will be a single rotation equal to the sum or difference of the components, except they are equal in magnitude but opposite in direction.

The latter case may be termed a couple of rotations. The result of such a couple is merely one motion of translation in the direction perpendicular to the plane of the couple, and measured by its moment, that is by the product of one of the rotations multiplied by the distance between their parallel axes.

The motion of any system about a fixed point is generally considered as a series of simple rotations about a certain axis, passing through that point, but continually changing its position; it being supposed to remain at rest for an instant, whence it is termed the instantaneous axis of rotation. This definition, though exact, requires the following additional explanation, in order to give a distinct idea of such motion.

The rotation of a body on an axis which continually changes its position about a fixed point, is the same as the motion of a cone with its vertex at that point, and rolling on the surface of another fixed cone with the same vertex. The varying line of contact of these two cones represents the instantaneous axis, which therefore moves within the body and in absolute space. If the motion of the body is given these two cones may be found, and *vice versa.*

Thus, in order to produce the exact rotatory motion of the earth, the circumference of the base of the rolling cone must have the same ratio to that of the base of the fixed cone, as a day to the time of a complete revolution of the equinoxes.

If a body has two motions at the same time, one of translation, the other of rotation, each may be considered separately, but the most distinct idea of its movements will be obtained by comparing the complex motion with that of a solid screw turning within its hollow companion.

With regard to the forces producing motion, it is obvious that a force applied in a direction passing through the centre of gravity of a body, will effect a motion of translation with a velocity measured by this force divided by the mass of the body; and that a couple of forces applied to a body in a plane, perpendicular to one of its three principal axes, will cause its rotation on this axis with an angular velocity equal to the moment of the couple divided by the moment of inertia about this axis.

The three components of a couple are three rotations directly proportional to the couples generating the latter, and inversely proportional to the moments of inertia about the three axes. The axis of the resulting rotation coincides, therefore, in direction with a diameter conjugate to the plane of the couple in an ellipsoid with axes inversely proportional to the square roots of the moments of inertia of the body about the same axes. The construction of such an ellipsoid about the centre of gravity of the body, and upon its three principal axes, is always conceivable, and being of the greatest importance in the theory of rotation, it has received the name — the central ellipsoid.

Supposing, now, that a body at rest is solicited by a couple of forces, in any plane passing through its centre of gravity, this plane being considered a diametrical plane of the central ellipsoid, we see that the instantaneous axis to which this couple gives rise, is the diameter conjugate to the plane of this couple in the ellipsoid, and it is evident that the angular velocity will be measured by the moment of the couple estimated perpendicular to this diameter, and divided by the moment of inertia of the body about the same diameter.

Since a couple may always be transferred in a plane parallel to its own, without altering its effect on the body, we may always assume that the plane of the couple of impulse, instead of passing through the centre, is a tangent plane to the surface of our ellipsoid.

If, therefore, a body is acted upon by a couple situated in any tangent plane of the central ellipsoid of this body, the instantaneous pole of the rotation to which this couple gives rise, is exactly at the point of contact.

Conversely, if a body rotates about any diameter of its central ellipsoid, the couple by which it is acted upon lies in the tangent plane passing through the pole.

The rotation about an instantaneous axis, which, by hypothesis, is not a principal axis of the rotating body, generates centrifugal forces in its particles which are not in equilibrium, and thus give rise to an accelerating couple. The effect of the latter is to impress the body at every instant with an infinitely small rotation, which is to be compounded with its actual rotation.

Taking two straight lines, the one of which represents the axis and magnitude of the couple of impulse, and the other the axis and magnitude of instantaneous rotation, then the accelerating couple due to the centrifugal forces will be represented in magnitude and direction by the surface of the parallelogram described upon the two lines.

The accelerating couple tends to turn the body about the diameter conjugate to the plane, passing through the axis of the couple of impulse and the instantaneous axis; this diameter is, therefore, at the same time conjugate to the latter and to its projection on the plane of the couple, whence we conclude that the axis about which the centrifugal forces tend to turn the body is in the same plane with the couple of impulse.

Taking two lines, one representing the infinitely small rotation, the other, the actual rotation, and completing the parallelogram, it becomes obvious that the end of its diagonal remains always at the same height above the plane of the couple.

It follows, therefore,—

1. During the whole course of motion the angular velocity is proportional to the radius drawn from the centre to the instantaneous pole, on the surface of the central ellipsoid.

2. The plane of the couple considered as tangent through the pole, remains always at the same distance from the centre of this ellipsoid.

This centre being fixed in absolute space, and the plane of the couple always remaining parallel to itself, therefore, this plane continually touching the central ellipsoid at the instantaneous pole of rotation, is always one and the same plane fixed in absolute space.

The central ellipsoid rolls without sliding on this plane with which it is continually in contact.

The series of points of contact considered upon the surface of the ellipsoid trace the motion of the instantaneous pole within the body; and the same points considered upon the fixed plane, trace its motion in absolute space. We can, therefore, determine these two curves, and may consider them as the bases of two conical surfaces with common vertex, one of which, movable with the body, rolls upon the other which is fixed in space.

In order to determine the first curve, it is only necessary to find the series of points where the ellipsoid would be touched by a movable plane which remains always at the same distance from the centre of the ellipsoid, or, in other words, which would touch at the same time this ellipsoid and a concentric sphere with a radius equal to the given distance.

It is obvious that this curve is an oval of double curvature, which, like the ellipse, has four vertices, where it is divided into four equal and symmetrical parts. These four vertices are the points where the radius vector and, consequently, the angular velocity of rotation attains its maxima and minima values. The maximum takes place when the instantaneous pole passes through the two vertices which lie in the mean principal section of the ellipsoid, and the minimum when it passes through the other two vertices of the curve.

The second curve may be considered as being generated by the former rolling about the centre upon the fixed plane of the couple. It is therefore a plane curve which goes round the projection of this centre in equal and regular wave lines corresponding to the equal and symmetrical arcs of the generating oval. It is a kind of circular curve, but with a variable radius, and which undulates between two concentric circles, touching alternately the one and the other.

If the angle at the centre corresponding to two consecutive summits of the equidistant waves is commensurable with four right angles, the curve is limited in all directions after a certain number of revolutions, and the instantaneous pole by which it is described, returns to the same point in the body as well as in absolute space. But, in the contrary case, the curve is infinite, and the pole which passes periodically through the same point in the body, cannot simultaneously return to the same point in absolute space.

The polar equations to these two curves, described at the same time by the same point, are identical.

The rolling cone, the surface of which has the first curve for a base, is merely a right cone of the second degree; but the fixed cone upon which it rolls, is a transcendental cone, the surface of which undulates about the fixed axis of the couple; it is also a kind of a right and circular cone, but with a grooved surface.

The compound term *polodia* will be an appropriate name for either of these curves; the one described within the body may be called the relative polodia, the other described in space, the absolute polodia; or, in order to distinguish them by their particular form, the first may be simply termed a polodia, the second a herpolodia.

These two curves depend on four given quantities, viz. the three principal radii of the central ellipsoid which are given by the nature of the body, and the height of the centre above the tangent plane of the couple which is given by the direction of the couple of impulse.

In the particular case where this height is equal to the length of the mean radius of the ellipsoid, the polodia becomes an ellipse, the plane of which passes through this mean radius; and the herpolodia becomes a spiral, with two branches turning in contrary directions about the same fixed centre. In this particular case of the motion of bodies, the instantaneous pole of rotation is always a new point in the body as well as in absolute space, though the length of the spiral is finite and equal to half the circumference of the rolling ellipse by which it is generated.

If the distance of the centre from the tangent plane is equal to one of the two extreme radii of the ellipsoid, which can only happen on a single point of the surface, the polodia and herpolodia are both reduced to one and the same point, and the instantaneous pole remains immovable in the body and in space during the whole course of the motion.

The same takes place when the plane of the couple is a tangent exactly at the point which is the mean pole of the central ellipsoid of the body.

If the body belongs to the species of those which have two of their principal moments of inertia equal to each other, in which case the central ellipsoid is one of revolution, the polodia becomes a circle about the axis of this spheroid, and the herpolodia is another circle about the fixed axis of the couple. In all bodies of this kind the motion is that of a right cone with circular base rolling upon another fixed cone likewise right and circular. If the circumferences of the two circles are not commensurable, as is generally the case, the instantaneous pole can never return to a point which coincides at once with the same point in the body and with the same point in absolute space.

The most simple case of all is that where the central ellipsoid of the body is a perfect sphere; for, however the couple may be applied, the axes of rotation and of the couple coincide, and the instantaneous pole remains immovable in the body and in absolute space.

Having examined the nature of the two curves which are described by the pole and the differential equations of which are identical when referred to the arc and radius vector emanating from the centre of the ellipsoid, we may consider the velocity with which the instantaneous pole describes at the same time the one curve and the other, and the velocity which this point has itself in moving from or approaching to the centre. This is the fluxion of the radius vector, and, consequently, of the angular velocity of rotation. We may likewise find the angular motion of the pole about the fixed axis of the couple of impulse. It is, therefore, easy to determine the remarkable points where these different velocities pass through their maximum or minimum. This will happen at the alternate summits of the waves of the herpolodia.

After that we may simplify the equation to the herpolodia, in referring it to a radius vector emanating from its proper centre taken in the plane of the curve, and to an angle which this radius describes about the same centre.

If we want to find the formulæ for calculating the place of the body at the end of any given time, we must express the angular velocity of rotation in functions of the time which will be effected by a first integration, and will give the place of the instantaneous pole at the surface of the central ellipsoid. Afterwards we integrate the preceding equation of the herpolodia, which will give the place of the pole on the fixed plane of the couple. By means of these two integrations, the proposed problem is entirely solved; that is, we are enabled to assign the place in space where the body ought to be at the end of any given time. We need only suppose the ellipsoid brought into contact with the fixed plane, so that it touches it at the point which we have determined on its surface and at the point which we have determined on the fixed plane, then by this operation the central ellipsoid and consequently the body itself will be found exactly in that position in absolute space to which it is carried by its proper motion at the end of the given time.

We may vary these determinations in many ways, taking other variables relatively to the position of the body; but whatever be the co-ordinates employed, the expression for these quantities in functions of the time will always require two integrations which essentially depend on elliptic transcendentals.

In the particular case where the height of the centre above the fixed plane is equal to the mean radius of the ellipsoid, and where in consequence thereof the polodia becomes a simple ellipse and the herpolodia a spiral, the difficulty disappears, the whole process is reduced to the ordinary rules depending merely on the integration of exponential and logarithmic functions.

Finally for all bodies where the central ellipsoid is one of revolution, and where, therefore, all becomes uniform and circular, no integration is necessary for determining the place of the body at the end of any given time.

When the plane of the couple of impulse is so situated that the instantaneous pole coincides with one of the principal poles of the central ellipsoid, it will remain stationary, so that the instantaneous axis, the axis of the body and that of the couple, always coincide in one and the same straight line immovable during the whole course of the motion. The principal axes of the body are therefore, all three, permanent axes of rotation; but there is a remarkable difference between them with regard to the stability which each of these axes may show in the rotation of the body.

If the instantaneous pole coincides with the major or minor pole of the ellipsoid, and is then removed from it a little distance by the impulse of a small extraneous couple, it will not easily move farther, it will describe its polodia about the same pole from which it has been removed.

But it is otherwise when the instantaneous pole coincides with the mean pole of the ellipsoid, for the slightest derangement will remove it farther, describing its polodia, either about the major pole or about the minor pole, according as this accidental derangement of the instantaneous axis, causes to increase or diminish the distance of the tangent plane of the couple from the centre of the ellipsoid. If the derangement is such that this distance does not vary, which happens on the surface along the two particular ellipses intersecting at the mean pole, then the instantaneous pole will describe the particular ellipse upon which it has been moved, or rather the half of this ellipse till it reaches the mean pole opposite, which is the greatest derangement that can happen to the body; now if the instantaneous pole has been carried to the other half of the same ellipse, it will immediately return to the mean pole whence it started, which is the least derangement possible of the rotation of the body.

There exists an exceptional case where the instantaneous axis being removed from the mean axis where it originally was, comes immediately nearer to it instead of moving farther; but in all the other cases it will describe an elliptic cone, either about the major or the minor axis, or trace the plane of one or the other of the particular ellipses intersecting at the mean pole, and we infer, therefore, that the rotation of the body about its mean axis is unstable.

The rotation is only stable about the axis of the greatest or least moment of inertia of the body, but we cannot herefrom conclude that it is equally stable for these two axes, because if one of them differs only little from the mean axis, it will not have much more stability than the latter.

In order to obtain a distinct idea of this stability, and of what constitutes its measure for each of the two extreme axes of the ellipsoid, let us imagine the surface of the latter to be divided into four parts or elliptic lunes, by means of the two ellipses, the planes of which intersect in the mean axis. The mean poles are therefore situated at the intersection of these two ellipses, the major poles are in the centres of two lunes, and the minor poles in the centres of the two supplementary lunes.

Now if the instantaneous pole of rotation coincides with

the mean pole of the ellipsoid, it is obvious that in consequence of a slight displacement, it will fall into one or the other of the two lunes in question, and describe its polodia about the one or the other principal pole of the ellipsoid. Further, if the displacement of the instantaneous pole takes place upon one of the two ellipses themselves, it will describe the half of this ellipse till it reaches the mean pole opposite, and being thus carried to the other half of the same ellipse it will immediately return to the mean pole whence it started. The result, therefore, is, as stated before, that the mean axis has no stability.

If the instantaneous pole coincides with the major pole of the ellipsoid, any displacement within the extent of the adjacent lune will not alter its path round the same major pole; and since this constitutes the stability of the major axis, we may say that the magnitude of that lune is in some way the measure of it. Similarly the surface of the supplementary lune is the measure of the stability of rotation about the minor axis.

Now if one of these two axes differs little from the mean axis, the lune corresponding to it will be very small, and its supplementary lune very great, therefore the axis slightly differing from the mean axis has very little stability, while the other possesses it in a high degree.

It is therefore not accurate to say, as is generally done, that if the instantaneous axis is a little removed from the principal axis which corresponds to the least or greatest moment of inertia of the body, it will move very little farther and only make small oscillations during the whole course of the motion; because if the moment of inertia relative to this axis differs little from the mean moment, the instantaneous pole can in consequence of a slight derangement, leave its small lune to enter the adjacent lune, there describing its polodia about the other axis; or even if the displacement is limited within the small lune, it can describe there a narrow and very elongated polodia, and therefore make very great oscillations about the principal pole from which it has been removed.

In those bodies where one of the extreme moments of inertia differs little from the mean moment, and where, therefore, the central ellipsoid of the body is nearly one of revolution about one of its axes, the stability of the rotation is only true for this axis.

This is the case with the earth, the rotation of which is very stable about its actual axis; but there would be very little stability in a rotation about the third axis, which, as we know, differs very little from the mean axis.

We may now seek to determine the motions of the poles of the central ellipsoid themselves, the velocity with which they turn round the fixed axis of the couple of impulse, and the velocity with which they sink and rise relatively to the plane perpendicular to this axis, by means of which they receive their movements of precession and nutation. We may also investigate the nature and properties of the three curves which the projections of these three principal poles trace in the same time upon the fixed plane, and we shall find without difficulty, many remarkable theorems on the motion of bodies, of which the following are some examples:—

If we draw from the centre upon the three principal axes of the body three equal lines, we shall find that the sum of the three variable areas described by the projections of these equal lines on the plane of the couple is proportional to the time.

If instead of taking three equal lines on the principal axes we take on them three lines proportional to the square roots of the moments of inertia, we shall find that the sum of the areas traced by the projections of these unequal lines is also proportional to the time.

If we consider the three principal poles of the central ellipsoid, we shall find that the sum of the squares of their distances from the axis of the couple of impulse is constant;

And that the sum of these squares multiplied respectively by the moments of inertia of the body is also constant during the whole course of the motion.

If we consider the curves generated by the projections of the three principal poles upon the fixed plane, we shall find that these curves are of the same species as the herpolodia, traced by the instantaneous pole of rotation.

In the general case one of the two poles, major or minor, which forms the centre of the polodia, describes a curve with equal and regular undulations about the same centre. The superior summits of the one correspond to the superior of the other, and the inferior to the inferior. Meanwhile the two other poles describe also curves regularly undulated, but so that when the one passes the highest points, the other passes the lowest points in the waves of its curve.

In the particular case where the polodia is an ellipse, and the herpolodia a spiral, the three poles of the body describe also spirals. There exists in this case a certain diameter within the mean plane of the central ellipsoid, which always remains perpendicular to the fixed axis of the couple, describing the plane of the latter with a uniform motion, so that the whole movement of the body consists in turning on this particular diameter with a variable velocity, while this diameter itself describes uniformly a circle in absolute space.

If the central ellipsoid is one of revolution, the pole of the figure will describe a circle like the instantaneous pole. Taking arbitrarily two other poles on the equator separated by a quadrant, their projections will describe two equal herpolodias with a common centre.

Finally, the general motion of a body may be represented by that of an elliptic cone, rolling upon the plane of the couple of impulse with a variable velocity and sliding upon it with a uniform velocity.

The principles just laid down must be well understood in order to obtain clear and distinct ideas, even of the most elementary parts of rotatory motion, but their greatest advantage consists in affording easy demonstrations for the movements of precession and nutation, thus simplifying and sometimes rectifying the most difficult theories.

NOTE.—Page 167.

TO DERIVE EXPRESSIONS FOR THE MOTION OF FALLING AND PROJECTED BODIES FROM THE DIFFERENTIAL EQUATIONS FOR THE MOTION OF THE SIMPLE PENDULUM.

Since, in the case of descent or projection, the accelerating force arising from the suspension of the pendulum vanishes, we substitute

$$\mu = o$$

in the equations (C), page 167; thus we get for the motion of free moving bodies, with regard to the rotation of the earth, the following—

$$\left.\begin{aligned} \frac{d^2\xi}{dt^2} &= -2\,n\sin\phi\,\frac{d\eta}{dt} \\ \frac{d^2\eta}{dt^2} &= 2\,n\sin\phi\,\frac{d\xi}{dt} - 2\cos\phi\,\frac{d\zeta}{dt} \\ \frac{d^2\zeta}{dt^2} &= 2\,n\cos\phi\,\frac{d\eta}{dt} + g \end{aligned}\right\} \text{(1.)}$$

The first integration gives

$$\left.\begin{aligned} \frac{d\xi}{dt} &= -2\,n\eta\sin\phi + \kappa \\ \frac{d\eta}{dt} &= 2\,n\,\xi\sin\phi - 2\,n\zeta\cos\phi + \kappa' \\ \frac{d\zeta}{dt} &= 2\,n\eta\cos\phi + gt + \kappa'' \end{aligned}\right\} \text{(2.)}$$

where κ, κ', κ'' denote the arbitrary constants. Multiplying the first of these equations by $\sin\phi$, the third by $(-\cos\phi)$ we have, by adding the products,

$$\frac{d\xi}{dt}\sin\phi - \frac{d\zeta}{dt}\cos\phi = -2\,n\,\eta - gt\cos\phi + 2\,nq$$

where

$$2nq=\kappa \sin\phi-\kappa'' \cos\phi$$

Substituting this value in the second equation of (1.), we obtain

$$\frac{d^2\eta}{dt^2}=-4n^2\eta-2ngt\cos\phi+4n^2q$$

which gives by integration

$$\eta=q+q' \cos 2nt+q'' \sin 2nt-\frac{gt}{2n}\cos\phi \quad . \quad (3.)$$

Substituting the first differential of this expression in the second equation of (2.), there results

$$\xi \sin\phi-\zeta \cos\phi=h-q' \sin 2nt+q'' \cos 2nt \quad . \quad (4.)$$

where

$$h=-\frac{g}{4n^2}\cos\phi-\frac{\kappa'}{2n}$$

Multiplying the first equation of (2.) by $\cos\phi$, the third by $\sin\phi$, then the integration of the sum of these products gives—

$$\xi \cos\phi+\zeta \sin\phi=c+c't+\tfrac{1}{2}gt^2 \sin\phi \quad . \quad . \quad (5.)$$

where c, the new arbitrary constant, and

$$c'=\kappa \cos\phi+\kappa'' \sin\phi$$

From the equations (4) and (5) we derive

$$\xi=p+c't \cos\phi-q' \sin\phi \sin 2nt$$
$$+q'' \sin\phi \cos 2nt+\tfrac{1}{2}gt^2 \sin\phi \cos\phi \quad . \quad (6.)$$
$$\zeta=p'+c't \sin\phi+q' \cos\phi \sin 2nt$$
$$-q'' \cos\phi \cos 2nt+\tfrac{1}{2}gt^2 \sin^2\phi \quad . \quad . \quad (7.)$$

where

$$p=\ \ h \sin\phi+c \cos\phi$$
$$p'=-h \cos\phi+c \sin\phi$$

The equations (3, 6 and 7) are the required integrals containing the six arbitrary constants q, q', q'', p, p', c', three of which are determined by taking $\xi=o$, $\eta=o$, $\zeta=o$ for $t=o$; thus we obtain

$$p=-q'' \sin\phi\,;\ p'=q'' \cos\phi\,;\ q=-q'$$

and the integrals are transformed into

$$\left.\begin{array}{l}\xi = c't\cos\phi - q'\sin\phi\sin 2nt - q''\sin\phi\,(1-\cos 2nt) \\ \qquad + \frac{1}{2}gt^2\sin\phi\cos\phi \\ \eta = -q'\,(1-\cos 2nt) + q''\sin 2nt - \frac{gt}{2n}\cos\phi \\ \zeta = c't\sin\phi + q'\cos\phi\sin 2nt + q''\cos\phi\,(1-\cos 2nt) \\ \qquad + \frac{1}{2}gt^2\sin^2\phi\end{array}\right\} \quad (8.)$$

The origin of co-ordinates is by this determination transferred to the point where the motion of the body commences.

We must now distinguish the two cases for falling and projected bodies. In the first case the initial velocity of the body is zero, whence

$$c' = o;\; q' = o;\; q'' = \frac{g}{4n^2}\cos\phi$$

and the integrals are, therefore,

$$\xi = -\frac{g}{4n^2}\sin\phi\cos\phi\,(1-\cos 2nt) + \frac{1}{2}gt^2\sin\phi\cos\phi$$

$$\eta = \frac{g}{4n^2}\cos\phi\sin 2nt - \frac{gt}{2n}\cos\phi$$

$$\zeta = \frac{g}{4n^2}\cos^2\phi\,(1-\cos 2nt) + \frac{1}{2}gt^2\sin^2\phi$$

The heights at our disposal for experiments on falling bodies are comparatively so small that without an appreciable error

$$\sin 2nt = 2nt - \frac{4}{3}n^3t^3$$
$$1-\cos 2nt = 2n^2t^2 - \frac{2}{3}n^4t^4$$

and the expressions for the co-ordinates become, therefore,

$$\left.\begin{array}{l}\xi = \frac{1}{6}gn^2t^4\sin\phi\cos\phi \\ \eta = -\frac{1}{3}gnt^3\cos\phi \\ \zeta = \frac{1}{2}gt^2 - \frac{1}{6}gn^2t^4\cos^2\phi\end{array}\right\} \quad (9.)$$

which are identical with Gauss' *Formulæ* (page 163) and show that the deviation towards the south is not sensible, while a sensible deviation takes place towards the east from the vertical of the original point of descent.

In the second case, where the body is projected, let us denote by a the initial azimuth of the direction of the motion, and by ν and ν' respectively, the horizontal and vertical components of the initial velocity, then

$$\nu \cos a = c' \cos\phi - 2nq' \sin\phi$$

$$\nu \sin a = 2nq'' - \frac{g}{2n}\cos\phi$$

$$-\nu' = c' \sin\phi + 2nq' \cos\phi$$

whence

$$c' = \nu \cos a \cos\phi - \nu' \sin\phi$$

$$q' = \frac{\nu}{2n}\cos a \sin\phi + \frac{\nu'}{2n}\cos\phi$$

$$q'' = \frac{\nu}{2n}\sin a + \frac{g}{4n^2}\cos\phi.$$

Substituting these values in the equations (8.), we get

$$\begin{aligned}
\xi = {} & \nu t \cos a \cos^2\phi - \nu' t \sin\phi \cos\phi \\
& + \frac{\nu}{2n}\cos a \sin^2\phi \sin 2nt + \frac{\nu'}{2n}\sin\phi\cos\phi \sin 2nt \\
& - \frac{\nu}{2n}\sin a \sin\phi\,(1-\cos 2nt) - \frac{g}{4n^2}\sin\phi\cos\phi\,(1-\cos 2nt) \\
& \quad + \tfrac{1}{2} g t^2 \sin\phi\cos\phi
\end{aligned}$$

$$\begin{aligned}
\eta = {} & \frac{\nu}{2n}\cos a \sin\phi\,(1-\cos 2nt) + \frac{\nu'}{2n}\cos\phi\,(1-\cos 2nt) \\
& + \frac{\nu}{2n}\sin a \sin 2nt + \frac{g}{4n^2}\cos\phi \sin 2nt - \frac{gt}{2n}\cos\phi
\end{aligned}$$

$$\begin{aligned}
\zeta = {} & \nu t \cos a \sin\phi\cos\phi - \nu' t \sin^2\phi \\
& - \frac{\nu}{2n}\cos a \sin\phi\cos\phi \sin 2nt - \frac{\nu'}{2n}\cos^2\phi \sin 2nt \\
& + \frac{\nu}{2n}\sin a \cos\phi\,(1-\cos 2nt) + \frac{g}{4n^2}\cos^2\phi\,(1-\cos 2nt) \\
& \quad + \tfrac{1}{2} g t^2 \sin^2\phi.
\end{aligned}$$

Sufficiently accurate values for $\sin 2nt$, and $\cos 2nt$, are obtained by taking the first terms in their respective expansions, and substituting them, there results

$$\xi = \nu t \cos \alpha - \nu n t^2 \sin \phi \sin \alpha$$
$$\eta = \nu t \sin \alpha + \nu n t^2 \sin \phi \cos \alpha + \nu' n t^2 \cos \phi$$
$$\zeta = -\nu' t + \tfrac{1}{2} g t^2 + \nu n t^2 \cos \phi \sin \alpha.$$

Taking the vertical component of the initial velocity

$$\nu' = o$$

the expressions for ξ and η may be written

$$\xi = \nu t \cos (\alpha + n t \sin \phi)$$
$$\eta = \nu t \sin (\alpha + n t \sin \phi)$$

whence we infer, that in consequence of the earth's rotation, horizontally projected bodies always deviate from the vertical plane of their initial direction towards the right-hand side of the observer.

Taking the horizontal component of the initial velocity

$$\nu = o$$

the expressions for the coordinates become

$$\xi = o$$
$$\eta = \nu' n t^2 \cos \phi$$
$$\zeta = -\nu' t + \tfrac{1}{2} g t^2.$$

This value of ζ shows that the projectile, after the lapse of the time

$$\tau = \frac{2 \nu'}{g}$$

returns to the horizontal plane from which it was thrown; for this moment, therefore, we have

$$\eta = \frac{4 \nu'^3 n}{g^2} \cos \phi$$

whence we infer, that in consequence of be earth's rotation, projected bodies fall back to a point which lies to the west from the original point of projection.

The respective deviations of projected and falling bodies take, therefore, place in contrary directions, and those of the former far exceed in magnitude those of the latter. Their ratio may be determined by substituting in the formula just found, and in the second of (9) the values

$$v' = \sqrt{2gs}; \quad t = \sqrt{\frac{2s}{g}}$$

where s denotes the greatest height of the ascending or descending body. Thus we obtain respectively

$$\eta = 8s \sqrt{\frac{2s}{g}}\, n \cos\phi$$

$$\eta = -\frac{2}{3}s \sqrt{\frac{2s}{g}}\, n \cos\phi.$$

Whence for the same heights and latitudes, the westerly deflection of a body projected vertically upwards is twelve times greater than the easterly deflection of a falling body.

LONDON
PRINTED BY SPOTTISWOODE AND CO.
NEW-STREET SQUARE

www.ingramcontent.com/pod-product-compliance
Lightning Source LLC
Chambersburg PA
CBHW021506210326
41599CB00012B/1151
* 9 7 8 3 3 3 7 0 0 3 0 4 3 *